Tracking the Deep Biosphere through Time

Tracking the Deep Biosphere through Time

Editors

Henrik Drake
Magnus Ivarsson
Christine Heim

MDPI • Basel • Beijing • Wuhan • Barcelona • Belgrade • Manchester • Tokyo • Cluj • Tianjin

Editors
Henrik Drake
Linnaeus University
Sweden

Magnus Ivarsson
Swedish Museum of Natural History
Sweden

Christine Heim
University of Cologne
Germany

Editorial Office
MDPI
St. Alban-Anlage 66
4052 Basel, Switzerland

This is a reprint of articles from the Special Issue published online in the open access journal *Geosciences* (ISSN 2076-3263) (available at: https://www.mdpi.com/journal/geosciences/special_issues/tracking_deep_biosphere).

For citation purposes, cite each article independently as indicated on the article page online and as indicated below:

LastName, A.A.; LastName, B.B.; LastName, C.C. Article Title. *Journal Name* **Year**, *Volume Number*, Page Range.

ISBN 978-3-03943-951-5 (Hbk)
ISBN 978-3-03943-952-2 (PDF)

Cover image courtesy of Magnus Ivarsson.

Contents

About the Editors

Henrik Drake, Associate Professor, Ph.D., is a geochemist with a research focus on ancient biosignatures in the deep biosphere.

Magnus Ivarsson, Associate Professor, is a paleobiologist with a research focus on the fossil record of deep igneous rock and early evolution of the deep biosphere.

Christine Heim, Professor of Geobiology, conducts research with a focus on recent and ancient biosignatures in extreme environments, especially the continental deep biosphere.

Editorial

Tracking the Deep Biosphere through Time

Henrik Drake [1,*], Magnus Ivarsson [2] and Christine Heim [3]

[1] Department of Biology and Environmental Science, Linnaeus University, 39182 Kalmar, Sweden
[2] Department of Paleobiology, Swedish Museum of Natural History, Box 50007, 10405 Stockholm, Sweden; magnus.ivarsson@nrm.se
[3] Institute for Geology and Mineralogy, University of Cologne, 50674 Cologne, Germany; christine.heim@uni-koeln.de
* Correspondence: henrik.drake@lnu.se

Received: 6 November 2020; Accepted: 13 November 2020; Published: 15 November 2020

Abstract: The oceanic and continental lithosphere constitutes Earth's largest microbial habitat, yet it is scarcely investigated and not well understood. The physical and chemical properties here are distinctly different from the overlaying soils and the hydrosphere, which greatly impact the microbial communities and associated geobiological and geochemical processes. Fluid–rock interactions are key processes for microbial colonization and persistence in a nutrient-poor and extreme environment. Investigations during recent years have spotted microbial processes, stable isotope variations, and species that are unique to the subsurface crust. Recent advances in geochronology have enabled the direct dating of minerals formed in response to microbial activity, which in turn have led to an increased understanding of the evolution of the deep biosphere in (deep) time. Similarly, the preservation of isotopic signatures, as well as organic compounds within fossilized micro-colonies or related mineral assemblages in voids, cements, and fractures/veins in the upper crust, provides an archive that can be tapped for knowledge about ancient microbial activity, including both prokaryotic and eukaryotic life. This knowledge sheds light on how lifeforms have evolved in the energy-poor subsurface, but also contributes to the understanding of the boundaries of life on Earth, of early life when the surface was not habitable, and of the preservation of signatures of ancient life, which may have astrobiological implications. The Special Issue "Tracking the Deep Biosphere through Time" presents a collection of scientific contributions that provide a sample of forefront research in this field. The contributions involve a range of case studies of deep ancient life in continental and oceanic settings, of microbial diversity in sub-seafloor environments, of isolation of calcifying bacteria as well as reviews of clay mineralization of fungal biofilms and of the carbon isotope records of the deep biosphere.

Keywords: deep biosphere; geobiology; deep time; geochronology; microorganisms; evolution

---

1. Introduction

The deep biosphere is estimated to represent a significant portion of all live biomass on Earth [1,2]. Its realm ranges downwards into sediments, sedimentary and igneous rock [3–5]. The microbial communities here include mainly prokaryotes (bacteria and archaea) [6] but findings of eukaryotes such as active and fossilized fungi [7–15], as well as live and indigenous nematodes [16], have also been reported at great depth. The strict energy limitation, absence of sunlight and dominantly anoxic conditions of these systems make them different to other ecosystems on Earth in terms of metabolic flexibility and recycling [4,17–22]. The deep biosphere is the second largest reservoir of live biomass today, only surpassed by land plants. However, it has been put forward that most of Earth's live biomass (~80%) was to be found in the deep biosphere prior to plant colonization of land [23]. The deep biosphere may thus have played a crucial role in the early evolution of both prokaryotes and eukaryotes, and dominated life on Earth for most of the planet's history. Yet, in addition to being an understudied environment in general when it comes to active communities, research on the deep time perspective of the deep biosphere has been particularly neglected. Microbiological and geobiological studies of the deep biosphere have, however, been the subject of increased interest in the last decade [10,18,24,25]. Similarly, the environmental conditions of these deep and more or less isolated settings have gained more attention lately [26]. By initiating the Special Issue, "Tracking the Deep Biosphere through Time", we aim to gather a collection of scientific contributions that provide a sample of forefront research in this emerging field of research. The contributions range from detection of ancient biosignatures in continental and oceanic settings, to microbial diversity in sub-seafloor environments and to isolation of calcifying bacteria. Two articles timely review clay mineral fossilization of fungal biofilms and the carbon isotope records of the deep biosphere.

2. An Overview of the Special Issue and the Contributions

A total of seven articles were published within this Special Issue; of which, five are research articles and two are reviews. A short summary for each contribution is given below.

Tillberg et al. [27] re-evaluate previously described fungal fossils [13] from the Lockne meteorite crater (age: 458 Ma), Sweden, by providing new in-situ Rb/Sr geochronological constraints of secondary calcite–albite–feldspar mineralization in veins and cavities of the impact structure, together with the fungal fossils. This new geochronological technique has shown to be feasible for dating complex discretely zoned mineral assemblages in fractured rock [28–32] and yielded a 357 ± 7 Ma age for the Lockne mineral assemblage. The study concludes that fungal colonization took place at least 100 Myr after the impact event, thus long after the impact-induced hydrothermal activity ceased. Microscale stable isotope data of ^{13}C-enriched calcite are also presented and reveled that microbial methanogenesis had occurred within the fracture system as well. Thus, the Lockne fungal fossils are not related to the impact event but represent later colonization of fractures formed by the impact.

In a study from the deep borehole, COSC-1, drilled into the Silurian–Devonian Scandinavian Caledonide mountain range in Central Sweden, Drake et al. [33] present isotopic and geochronological constraints for ancient microbial life in deep fractures and micro-karst formations in the nappe system. Micro-karsts at 122–178 m depth featured several generations of secondary calcite and pyrite growth. The younger of these precipitation phases showed ^{34}S-depleted $\delta^{34}S_{pyrite}$ values consistent with microbial sulfate reduction in situ. Laser ablation inductively coupled mass spectrometry carbonate U-Pb geochronology [34] was applied to the late stage calcite of this assemblage and gave two separate ages (9.6 ± 1.3 Ma and 2.5 ± 0.2 Ma). These ages mark fluid circulation, related bacterial activity and mineral precipitation, following karst formation. The results show that the combined high spatial-resolution stable isotope and geochronology approach is suitable for characterizing paleo-fluid flow in micro-karst in nappe units.

Carlsson et al. [35] describe filamentous fossils interpreted as fungal remains from the Troodos ophiolite (91 Ma), Cyprus. Mineralogy and fluid inclusion studies show that the microorganisms lived in the system while the volcanism-related hydrothermal activity still was active in the seafloor, thus before the crust was emplaced on land. The paper highlights the importance of detailed geochemical investigations associated with descriptions of potential microfossils in the deep subsurface. Reconstructing the past habitats, fluid regimes and geochemical conditions are key in understanding fossilization and addressing biogenicity criteria.

Martino et al. [36] explore the microbial community of the sub-seafloor sediments obtained during IODP (International Ocean Discovery Program) Expedition 334 at Costa Rica Margin. The correlation assessment between microbial taxa (Bacteria, Archaea and Eukaryotes) and environmental chemistry surprisingly diverge despite their geographical proximity, similar lithology and similar pore water chemistry. The major taxonomic lineages identified have also been found in sub-seafloor environments worldwide. Up to now, knowledge about habitat preferences and metabolic specializations is rather low regarding this environment and therefore variations in their relative abundance are just partly explained. However, this study also shows that contamination issues for such subsurface drilling projects seem well manageable and therefore should not hinder further necessary research.

Cacchio and Gallo [37] provide a methodical study describing calcium-carbonate precipitation processes in bacterial strains isolated from the rhizosphere. In contrast to bacterial carbonate precipitation in speleothems, the biomineralization potential in rhizosphere soils is understudied. Assessing the carbonate precipitation in the original environment and in enrichment cultures, this is the first study to show the biomineralization ability of the bacterial strains Vibrio tubiashii, Clavibacter agropyri (Corynebacterium), Corynebacterium urealyticum, Sanguibacter suarezii, Sphingomonas sanguinis, and Pseudomonas syringae, which are commonly not related to mineral precipitation. Initial mineral precipitation was induced in direct vicinity of the bacterial cells and thus attributable to the metabolic activity of the cells. Besides presenting well adapted and versatile subsurface organisms, the authors highlight methodical approaches for efficient and cost-optimized microbial community analysis and rapid screenings.

Sallstedt et al. [38] review the current data on clay mineralization by fossil fungal biofilms from oceanic and continental subsurface igneous rock. The aim of the review is to compare the nature of subsurface fungal clays from different igneous settings to evaluate the importance of host rock and ambient redox conditions for clay speciation related to fossilization of microorganisms. Montmorillonite-like smectite is the dominant clay involved in subsurface microbial fossilization, independent of host rock type and geochemical regime. The paper highlights the role of fungal clay authigenesis in the cycling of elements between host rock, ocean and precipitation of secondary minerals.

Meister and Reyes [39] review the carbon isotope record of the sub-seafloor biosphere, which exhibits large carbon-isotope fractionation effects as a result of microbial enzymatic reactions. The review presents a brief and timely overview on carbon isotopes, including microbial fractionation mechanisms, transport effects, preservation in diagenetic carbonate archives, and their implications for the past sub-seafloor biosphere. Particularly large carbon-isotope fractionation linked to methanogenesis and fractionations occurring during this process is a focus of the review. Methane metabolisms in Archaea and Bacteria essentially mediate through the enzymatic Wood–Ljungdahl pathway, but it remains unclear where in the pathway carbon-isotope fractionation occurs. In marine porewaters (and diagenetic carbonates), the highly varying C isotopic signature of dissolved inorganic carbon and methane reflects different microbial metabolisms [40]. Carbon isotope compositions of diagenetic carbonates may therefore act as biosignatures for ancient deep biosphere conditions.

3. Summary and Outlook

As stated in the introduction, the deep biosphere is understudied, particularly in a deep time perspective. With this Special Issue, we have gathered a collection of well-needed additions to take the state of knowledge further, and that we hope can inspire new research on both fossil and active communities in this vast realm.

Author Contributions: H.D., M.I. and C.H. handled the Special Issue and wrote the editorial. All authors have read and agreed to the published version of the manuscript.

Funding: The research projects of the authors of this editorial are funded by Formas (contract 2017-00766 to H.D.) and the Swedish Research Council (contract 2017-05186 to H.D., 2017-04129 to M.I.).

Acknowledgments: We thank the authors contributing to this Special Issue and Daisy Hu at MDPI for handling.

Conflicts of Interest: The authors declare no conflict of interest.

References

1. McMahon, S.; Parnell, J. Weighing the deep continental biosphere. *FEMS Microbiol. Ecol.* **2014**, *87*, 113–120. [CrossRef]
2. Magnabosco, C.; Lin, L.H.; Dong, H.; Bomberg, M.; Ghiorse, W.; Stan-Lotter, H.; Pedersen, K.; Kieft, T.L.; van Heerden, E.; Onstott, T.C. The biomass and biodiversity of the continental subsurface. *Nat. Geosci.* **2018**, *11*, 707–717. [CrossRef]
3. Pedersen, K. Exploration of deep intraterrestial microbial life: Current perspectives. *FEMS Microbiol. Lett.* **2000**, *185*, 9–16. [CrossRef] [PubMed]
4. Hoehler, T.M.; Jorgensen, B.B. Microbial life under extreme energy limitation. *Nat. Rev. Genet.* **2013**, *11*, 83–94. [CrossRef] [PubMed]
5. D'Hondt, S.; Jørgensen, B.B.; Miller, D.J.; Batzke, A.; Blake, R.; Cragg, B.A.; Cypionka, H.; Dickens, G.R.; Ferdelman, T.; Hinrichs, K.-U.; et al. Distributions of Microbial Activities in Deep Subseafloor Sediments. *Science* **2004**, *306*, 2216. [CrossRef]
6. Lopez-Fernandez, M.; Simone, D.; Wu, X.; Soler, L.; Nilsson, E.; Holmfeldt, K.; Lantz, H.; Bertilsson, S.; Dopson, M. Metatranscriptomes Reveal That All Three Domains of Life Are Active but Are Dominated by Bacteria in the Fennoscandian Crystalline Granitic Continental Deep Biosphere. *mBio* **2018**, *9*, e01792. [CrossRef]
7. Drake, H.; Ivarsson, M.; Bengtson, S.; Heim, C.; Siljeström, S.; Whitehouse, M.J.; Broman, C.; Belivanova, V.; Åström, M.E. Anaerobic consortia of fungi and sulfate reducing bacteria in deep granite fractures. *Nat. Commun.* **2017**, *8*, 55. [CrossRef]
8. Ivarsson, M.; Kilias, S.P.; Broman, C.; Neubeck, A.; Drake, H.; Fru, E.C.; Bengtson, S.; Naden, J.; Detsi, K.; Whitehouse, M.J. Exceptional Preservation of Fungi as H2-Bearing Fluid Inclusions in an Early Quaternary Paleo-Hydrothermal System at Cape Vani, Milos, Greece. *Minerals* **2019**, *9*, 749. [CrossRef]
9. Drake, H.; Ivarsson, M. The role of anaerobic fungi in fundamental biogeochemical cycles in the deep biosphere. *Fungal Biol. Rev.* **2018**, *32*, 20–25. [CrossRef]
10. Ivarsson, M.; Bengtson, S.; Drake, H.; Francis, W. Fungi in Deep Subsurface Environments. In *Advances in Applied Microbiology*; Academic Press: London, UK, 2018; Volume 102, pp. 83–116.
11. Bengtson, S.; Rasmussen, B.; Ivarsson, M.; Muhling, J.; Broman, C.; Marone, F.; Stampanoni, M.; Bekker, A. Fungus-like mycelial fossils in 2.4-billion-year-old vesicular basalt. *Nat. Ecol. Evol.* **2017**, *1*, 0141. [CrossRef]
12. Ivarsson, M.; Bengtson, S.; Skogby, H.; Lazor, P.; Broman, C.; Belivanova, V.; Marone, F. A Fungal-Prokaryotic Consortium at the Basalt-Zeolite Interface in Subseafloor Igneous Crust. *PLoS ONE* **2015**, *10*, e0140106. [CrossRef] [PubMed]
13. Ivarsson, M.; Broman, C.; Sturkell, E.; Ormö, J.; Siljeström, S.; van Zuilen, M.; Bengtson, S. Fungal colonization of an Ordovician impact-induced hydrothermal system. *Sci. Rep.* **2013**, *3*, 3487. [CrossRef] [PubMed]
14. Purkamo, L.; Kietäväinen, R.; Miettinen, H.; Sohlberg, E.; Kukkonen, I.; Itävaara, M.; Bomberg, M. Diversity and functionality of archaeal, bacterial and fungal communities in deep Archaean bedrock groundwater. *FEMS Microbiol. Ecol.* **2018**, *94*, 94. [CrossRef] [PubMed]
15. Sohlberg, E.; Bomberg, M.; Miettinen, H.; Nyyssönen, M.; Salavirta, H.; Vikman, M.; Itävaara, M. Revealing the unexplored fungal communities in deep groundwater of crystalline bedrock fracture zones in Olkiluoto, Finland. *Front. Microbiol.* **2015**, *6*, 573. [CrossRef] [PubMed]
16. Borgonie, G.; García-Moyano, A.; Litthauer, D.; Bert, W.; Bester, A.; van Heerden, E.; Möller, C.; Erasmus, M.; Onstott, T.C. Nematoda from the terrestrial deep subsurface of South Africa. *Nature* **2011**, *474*, 79–82. [CrossRef]
17. Mhatre, S.S.; Kaufmann, S.; Marshall, I.P.G.; Obrochta, S.; Andrèn, T.; Jørgensen, B.B.; Lomstein, B.A. Microbial biomass turnover times and clues to cellular protein repair in energy-limited deep Baltic Sea sediments. *FEMS Microbiol. Ecol.* **2019**, *95*, 95. [CrossRef]
18. Li, J.; Mara, P.; Schubotz, F.; Sylvan, J.B.; Burgaud, G.; Klein, F.; Beaudoin, D.; Wee, S.Y.; Dick, H.J.B.; Lott, S.; et al. Recycling and metabolic flexibility dictate life in the lower oceanic crust. *Nature* **2020**, *579*, 250–255. [CrossRef]

19. Wu, X.; Holmfeldt, K.; Hubalek, V.; Lundin, D.; Åström, M.; Bertilsson, S.; Dopson, M. Microbial metagenomes from three aquifers in the Fennoscandian shield terrestrial deep biosphere reveal metabolic partitioning among populations. *ISME J.* **2016**, *10*, 1192–1203. [CrossRef]
20. Kieft, T.L.; Walters, C.C.; Higgins, M.B.; Mennito, A.S.; Clewett, C.F.M.; Heuer, V.; Pullin, M.J.; Hendrickson, S.; van Heerden, E.; Sherwood Lollar, B.; et al. Dissolved organic matter compositions in 0.6–3.4 km deep fracture waters, Kaapvaal Craton, South Africa. *Org. Geochem.* **2018**, *118*, 116–131. [CrossRef]
21. Lau, M.C.Y.; Kieft, T.L.; Kuloyo, O.; Linage-Alvarez, B.; van Heerden, E.; Lindsay, M.R.; Magnabosco, C.; Wang, W.; Wiggins, J.B.; Guo, L.; et al. An oligotrophic deep-subsurface community dependent on syntrophy is dominated by sulfur-driven autotrophic denitrifiers. *Proc. Natl. Acad. Sci. USA* **2016**, *113*, E7927–E7936. [CrossRef]
22. Lever, M.A.; Rogers, K.L.; Lloyd, K.G.; Overmann, J.; Schink, B.; Thauer, R.K.; Hoehler, T.M.; Jørgensen, B.B. Life under extreme energy limitation: A synthesis of laboratory- and field-based investigations. *FEMS Microbiol. Rev.* **2015**, *39*, 688–728. [CrossRef] [PubMed]
23. McMahon, S.; Parnell, J. The deep history of Earth's biomass. *J. Geol. Soc.* **2018**, *175*, 716. [CrossRef]
24. Ivarsson, M.; Drake, H.; Neubeck, A.; Sallstedt, T.; Bengtson, S.; Roberts, N.M.W.; Rasmussen, B. The fossil record of igneous rock. *Earth-Sci. Rev.* **2020**, *210*, 103342. [CrossRef]
25. Inagaki, F.; Hinrichs, K.U.; Kubo, Y.; Bowles, M.W.; Heuer, V.B.; Hong, W.L.; Hoshino, T.; Ijiri, A.; Imachi, H.; Ito, M.; et al. Exploring deep microbial life in coal-bearing sediment down to ~2.5 km below the ocean floor. *Science* **2015**, *349*, 420. [CrossRef] [PubMed]
26. Lollar, G.S.; Warr, O.; Telling, J.; Osburn, M.R.; Lollar, B.S. 'Follow the Water': Hydrogeochemical Constraints on Microbial Investigations 2.4 km Below Surface at the Kidd Creek Deep Fluid and Deep Life Observatory. *Geomicrobiol. J.* **2019**, *36*, 859–872. [CrossRef]
27. Tillberg, M.; Ivarsson, M.; Drake, H.; Whitehouse, J.M.; Kooijman, E.; Schmitt, M. Re-Evaluating the Age of Deep Biosphere Fossils in the Lockne Impact Structure. *Geosciences* **2019**, *9*, 202. [CrossRef]
28. Tillberg, M.; Drake, H.; Zack, T.; Kooijman, E.; Whitehouse, M.J.; Åström, M.E. In situ Rb-Sr dating of slickenfibres in deep crystalline basement faults. *Sci. Rep.* **2020**, *10*, 562. [CrossRef]
29. Tillberg, M.; Maskenskaya, O.M.; Drake, H.; Hogmalm, J.K.; Broman, C.; Fallick, A.E.; Åström, M.E. Fractionation of Rare Earth Elements in Greisen and Hydrothermal Veins Related to A-Type Magmatism. *Geofluids* **2019**, *2019*, 20. [CrossRef]
30. Drake, H.; Whitehouse, M.J.; Heim, C.; Reiners, P.W.; Tillberg, M.; Hogmalm, K.J.; Dopson, M.; Broman, C.; Åström, M.E. Unprecedented ^{34}S-enrichment of pyrite formed following microbial sulfate reduction in fractured crystalline rocks. *Geobiology* **2018**, *16*, 556–574. [CrossRef]
31. Drake, H.; Ivarsson, M.; Tillberg, M.; Whitehouse, M.; Kooijman, E. Ancient Microbial Activity in Deep Hydraulically Conductive Fracture Zones within the Forsmark Target Area for Geological Nuclear Waste Disposal, Sweden. *Geosciences* **2018**, *8*, 211. [CrossRef]
32. Drake, H.; Heim, C.; Roberts, N.M.W.; Zack, T.; Tillberg, M.; Broman, C.; Ivarsson, M.; Whitehouse, M.J.; Åström, M.E. Isotopic evidence for microbial production and consumption of methane in the upper continental crust throughout the Phanerozoic eon. *Earth Planet. Sci. Lett.* **2017**, *470*, 108–118. [CrossRef]
33. Drake, H.; Roberts, M.W.N.; Whitehouse, J.M. Geochronology and Stable Isotope Analysis of Fracture-Fill and Karst Mineralization Reveal Sub-Surface Paleo-Fluid Flow and Microbial Activity of the COSC-1 Borehole, Scandinavian Caledonides. *Geosciences* **2020**, *10*, 56. [CrossRef]
34. Roberts, N.M.W.; Drost, K.; Horstwood, M.S.A.; Condon, D.J.; Chew, D.; Drake, H.; Milodowski, A.E.; McLean, N.M.; Smye, A.J.; Walker, R.J.; et al. Laser ablation inductively coupled plasma mass spectrometry (LA-ICP-MS) U–Pb carbonate geochronology: Strategies, progress, and limitations. *Geochronology* **2020**, *2*, 33–61. [CrossRef]
35. Carlsson, D.-T.; Ivarsson, M.; Neubeck, A. Fossilized Endolithic Microorganisms in Pillow Lavas from the Troodos Ophiolite, Cyprus. *Geosciences* **2019**, *9*, 456. [CrossRef]
36. Martino, A.; Rhodes, M.E.; León-Zayas, R.; Valente, I.E.; Biddle, J.F.; House, C.H. Microbial Diversity in Sub-Seafloor Sediments from the Costa Rica Margin. *Geosciences* **2019**, *9*, 218. [CrossRef]
37. Cacchio, P.; Del Gallo, M. A Novel Approach to Isolation and Screening of Calcifying Bacteria for Biotechnological Applications. *Geosciences* **2019**, *9*, 479. [CrossRef]

38. Sallstedt, T.; Ivarsson, M.; Drake, H.; Skogby, H. Instant Attraction: Clay Authigenesis in Fossil Fungal Biofilms. *Geosciences* **2019**, *9*, 369. [CrossRef]
39. Meister, P.; Reyes, C. The Carbon-Isotope Record of the Sub-Seafloor Biosphere. *Geosciences* **2019**, *9*, 507. [CrossRef]
40. Meister, P.; Liu, B.; Khalili, A.; Böttcher, M.E.; Jørgensen, B.B. Factors controlling the carbon isotope composition of dissolved inorganic carbon and methane in marine porewater: An evaluation by reaction-transport modelling. *J. Mar. Syst.* **2019**, *200*, 103227. [CrossRef]

Publisher's Note: MDPI stays neutral with regard to jurisdictional claims in published maps and institutional affiliations.

Article

Geochronology and Stable Isotope Analysis of Fracture-Fill and Karst Mineralization Reveal Sub-Surface Paleo-Fluid Flow and Microbial Activity of the COSC-1 Borehole, Scandinavian Caledonides

Henrik Drake [1],*, Nick M. W. Roberts [2] and Martin J. Whitehouse [3]

[1]   Department of Biology and Environmental Science, Linnæus University, 39182 Kalmar, Sweden
[2]   Geochronology and Tracers Facility, British Geological Survey, Nottingham NG12 5GG, UK; nirob@bgs.ac.uk
[3]   Swedish Museum of Natural History, 114 18 Stockholm, Sweden; martin.whitehouse@nrm.se
*   Correspondence: henrik.drake@lnu.se

Received: 13 December 2019; Accepted: 27 January 2020; Published: 3 February 2020

Abstract: The deep biosphere hosted in fractured rocks within the upper continental crust is one of the least understood and studied ecological realms on Earth. Scarce knowledge of ancient life and paleo-fluid flow within this realm is owing to the lack of deep drilling into the crust. Here we apply microscale high spatial-resolution analytical techniques to fine-grained secondary minerals in a deep borehole (COSC-1) drilled into the Silurian-Devonian Scandinavian Caledonide mountain range in central Sweden. The aim is to detect and date signs of ancient microbial activity and low-temperature fluid circulation in micro-karsts (foliation-parallel dissolution cavities in the rock) and fractures at depth in the nappe system. Vein carbonates sampled at 684 to 2210 m show a decreased C isotope variability at depths below 1050 m; likely due to decreased influence of organic-C at great depth. Micro-karsts at 122–178 m depth feature at least two generations of secondary calcite and pyrite growth in the voids as shown by secondary ion mass spectrometry analytical transects within individual grains. The younger of these two precipitation phases shows ^{34}S-depleted $\delta^{34}S_{pyrite}$ values ($-19.8 \pm 1.6‰$ vs. Vienna-Canyon Diablo Troilite (V-CDT)) suggesting microbial sulfate reduction in situ. The calcite of this late phase can be distinguished from the older calcite by higher $\delta^{18}O_{calcite}$ values that correspond to precipitation from ambient meteoric water. The late stage calcite gave two separate laser ablation inductively coupled mass spectrometry-derived U-Pb ages (9.6 ± 1.3 Ma and 2.5 ± 0.2 Ma), marking a minimum age for widespread micro-karst formation within the nappe. Several stages of fluid flow and mineral precipitation followed karst formation; with related bacterial activity as late as the Neogene-Quaternary; in structures presently water conducting. The results show that our combined high spatial-resolution stable isotope and geochronology approach is suitable for characterizing paleo-fluid flow in micro-karst; in this case, of the crystalline crust comprising orogenic nappe units.

Keywords: in situ U-Pb geochronology; secondary minerals; stable isotopes; Caledonides; deep drilling (COSC-1)

---

1. Introduction

Fluid circulation and mixing in fractures and during vein formation can lead to metal leaching and accumulation of ores, such as those in the Harz mountains, Germany [1]. Microbial activity in fractured rock volumes can also be of importance for ore-forming processes [2,3] and may involve potentially significant microbial natural gas accumulations [4,5]. Furthermore, microbial activity and fluid flow may involve important redox processes and fluctuations that are relevant for safety assessments of repositories for toxic wastes such as spent nuclear fuel [6,7]. Knowledge of ancient microbial processes and fluid flow in the upper crust can, therefore, have wide-ranging implications. Information about these processes can be preserved as diagnostic isotope signatures within authigenic minerals over geological timescales, for instance as excursions in $^{13}C/^{12}C$ ($\delta^{13}C$) in carbonate due to methane oxidation or formation [8–11] as well as excursions in $^{34}S/^{32}S$ ($\delta^{34}S$) in sulfides due to microbial sulfate reduction (MSR) [12–14]. These isotopic markers are related to the fractionation that occurs during microbial metabolisms, at degrees that are beyond what abiotic sources and thermochemical reactions produce [15,16]. In addition, the $^{18}O/^{16}O$ composition ($\delta^{18}O$) of the carbonate can reveal origins of fluids [6,17,18] owing to the fact that the O isotope composition is a conservative tracer for different water types [19], when temperature-related fractionation during calcite formation has been considered [20–22].

In deep granitoid fractures of the Fennoscandian Shield in eastern Sweden (Laxemar and Forsmark sites) and western Finland (Olkiluoto, Figure 1), several recent studies have applied micro-scale secondary ion mass spectrometry (SIMS) techniques to fine-grained secondary low-temperature calcite and pyrite mineral coatings, revealing several discrete events of microbial-related mineral precipitation [14,23–27]. The isotopic variability for both $\delta^{34}S$ (−54‰ to +132‰ V-CDT, Vienna-Canyon Diablo Troilite reference value) and $\delta^{13}C$ (−125‰ to +37‰ V-PDB, Vienna-Pee Dee Belemnite reference value) [14,28,29], is beyond what has been reported from other settings. At Forsmark, U-Pb carbonate geochronology using laser-ablation inductively coupled mass spectrometry (LA-ICP-MS) was successfully applied to calcite with anaerobic oxidation of methane-related (^{13}C-depleted) composition, revealing a Jurassic age [28]. In the fractured Devonian impact structure at Siljan, central Sweden, U-Pb dating was used to constrain dominantly Eocene–Miocene ages of secondary calcite formed following microbial methanogenesis and methane oxidation [5].

Here we use the aforementioned high spatial resolution isotopic and geochronological techniques to decipher low-temperature fluid circulation and to trace ancient microbial activity within a single deep borehole. The COSC-1 borehole is part of the Collisional Orogeny in the Scandinavian Caledonides (COSC) project, part-funded through the International Continental Drilling Project (ICDP). COSC focuses on mountain-building processes in a major mid-Paleozoic orogen in western Scandinavia and its comparison with modern analogues [30,31]. The deep borehole and core, with ~2.5 km at ~100% recovery, offers an excellent opportunity for geophysical and geochemical characterization of the crystalline crust. For the Scandinavian Caledonides, extensive characterization exists for tectonics of the nappe units [32,33] and for denudation and uplift [34,35], but knowledge of low-temperature fluid circulation and microbial activity in the deep fracture systems of the nappes is non-existent. Secondary mineral investigations of Caledonide-related hydrothermal brine-type mineralizations have been carried out, such as at Laisvall, Vassbo and Osen, but these mineralizations are temporally very close to the Caledonide orogenic event itself [36]. The preliminary borehole investigations of COSC-1 have shown that there are several micro-karst horizons formed within carbonate-rich layers of the gneisses, as well as abundant calcite veins [30,37,38], which both enable studies of low-temperature mineral formation and consequently of fluid circulation and ancient microbial activity. This study of secondary carbonate and sulfide mineralization in deep micro-karsts and veins of the COSC-1 core aims to increase our understanding of low-temperature fluid circulation and microbial activity in the nappe system of an ancient orogeny.

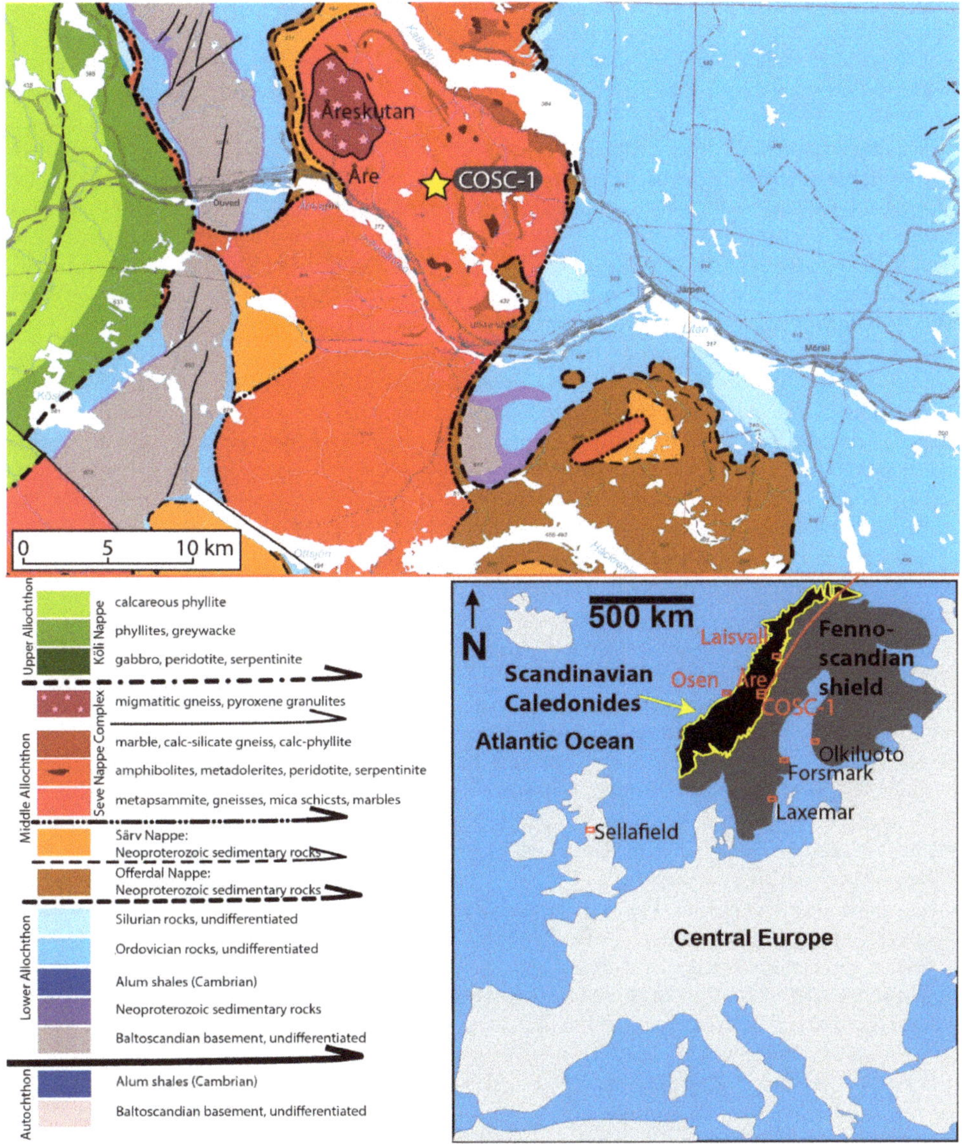

Figure 1. Geological map of the study area and the location of the COSC-1 (Collisional Orogeny in the Scandinavian Caledonides) borehole (yellow star symbol marks surface location, at coordinates 63°24′ N, 13°5′ E) along with information about the thrust sheets. Modified from Lorenz et al., [30,38]. Sites of previous geochemical and geobiological studies in fractures and veins are also indicated.

2. Geological Setting and the COSC-1 (Collisional Orogeny in the Scandinavian Caledonides) Borehole

The COSC-1 borehole was drilled in 2014 and is a ca. 2.5 km deep (2495.8 m), cored drill hole located close to the town of Åre in Jämtland, central Sweden (Figure 1). It targeted a thick section of the lower part of the Seve Nappe Complex and was planned to penetrate its basal thrust zone into the underlying lower-grade metamorphosed allochthon, but drilling did not reach that planned horizon [30].

The Caledonides of western Scandinavia and eastern Greenland formed as the result of a set of events that started with the closure of the Iapetus Ocean during the Ordovician, and subsequent underthrusting of continent Laurentia by Baltica in the Silurian and Early Devonian during the Scandian collisional orogenic stage. The allochthons were subjected to high-grade metamorphism and emplaced onto the adjacent platforms by eastward thrust emplacement by up to several hundred kilometres [33]. In central Sweden, the thrust sheets are divided into the Lower, Middle, Upper and Uppermost allochthons [39] that are unconformably overlying the Proterozoic crystalline basement, as marked by a front thrust sheet that dips 1–2° to the west [31]. Sedimentary successions of Neoproterozoic and Cambro-Silurian strata dominate the Lower Allochthon (Jämtlandian Nappes). The overlying Middle Allochthon is of higher metamorphic grade and contains a basal basement-derived thrust sheet, overlain by Offerdal Nappe metasandstones, and the Särv Nappe [40,41]. The Seve Nappe Complex is the uppermost tectonic unit in the Middle Allochthon and has a lower part that has experienced ductile deformation in dominantly amphibolite facies; a central part (e.g., Åreskutan Nappe) of migmatites and paragneisses and an upper, amphibolite-dominated unit [42]. Absolute ages reported from migmatite and associated rocks of the Åreskutan mountain 6 km to the west of the COSC-1 borehole are in the range of c. 455 to 420 Ma [43–47]. The Köli Nappes in the Upper Allochthon are the tectonostratigraphically highest rocks in study area (Figure 1). This unit is dominated by sedimentary rocks of Early Paleozoic age, and has experienced greenschist facies [42]. The Uppermost Allochthon contains metasedimentary rocks of inferred Laurentian margin origin, but are not present in the study area.

The drilling of the COSC-1 borehole aimed to study mountain building processes at mid-crustal levels in a major orogen, with a focus on the Seve Nappe Complex. The borehole lithology is briefly described in Figure S1 and, according to Lorenz et al., [30,38], the core has the following characteristics: the upper 1800 m is dominated by gneisses of varying compositions (e.g., felsic, amphibole, Calcium silicate) belonging to the Lower Seve Nappes. Highly strained metagabbros and amphibolites are common, and marbles, pegmatite dykes and minor mylonites also occur. Fractures that are interpreted to be of young, i.e., post-orogen, age are sparse. A water-conducting set of very steep fractures has resulted in dissolution of bands in the gneisses to form foliation parallel micro-karst (e.g., at about 175 m and between 1200 and 1320 m). Narrow deformation bands and mylonites mark the first signs of increasing strain below 1700 m. The base of the Seve nappes was interpreted to be at c. 2000–2100 m [37]. Below 2100 m, mylonites dominate. The lower part of the drill core is dominated by mylonitized quartzites and metasandstones of unclear tectonostratigraphic position that are mylonitized to varying degrees. A set of hydrological tests run during the drilling campaign revealed the location of likely water-conductive fractures between the tested depth range of 300–2500 m, these occurred at 339, 507, 554, 696, 1214, 1245, 2300 and 2380 m [48]. Borehole temperature profiling suggest a geothermal gradient of ~20 °C km^{-1} with temperatures reaching almost 55 °C in the bottom of the hole [30].

3. Materials and Methods

Eight samples containing secondary minerals (carbonate ± sulfides) were collected from the COSC-1 borehole core (core log with samples marked in Supplementary Figure S1). The samples were cut into thick polished blocks and analyzed with a petrographic microscope and scanning electron microscope (SEM). Micro-karst samples were analyzed directly in the drill core specimen and euhedral calcite and pyrite crystals were hand-picked from the karst voids and embedded in epoxy. These epoxy grain mounts were polished to expose a cross section of the crystals, and the interiors of these crystals were targeted with SIMS analysis for stable C, O, (calcite) and S (pyrite) isotopes and with LA-ICP-MS for U-Pb geochronology (calcite) after SEM-documentation of zonations and impurities.

3.1. Scanning Electron Microscopy (SEM)

The mineralogy and appearance of the uncoated fracture coatings and rock chips of veins were examined under low-vacuum conditions in a Hitachi S-3400N scanning electron microscope (SEM) equipped with an integrated energy-dispersive spectroscopy (EDS) system. The coatings were then scraped off for analyses of stable isotopes and U-Pb geochronology.

3.2. Secondary Ion Mass Spectrometry for $\delta^{13}C$, $\delta^{18}O$, $\delta^{34}S$

Intra-crystal SIMS-analysis (10 µm lateral beam dimension, 1–2 µm depth dimension) of sulfur isotopes in pyrite and carbon and oxygen isotopes in calcite was performed on a Cameca IMS1280 ion microprobe at the NordSIM facility at the Museum of Natural History, Stockholm, Sweden, following the analytical settings and tuning reported previously [14,29,49]. Sulfur was sputtered using a $^{133}Cs+$ primary beam with 20 kV incident energy (10 kV primary, −10 kV secondary) and a primary beam current of ~1.5 nA. A normal incidence electron gun was used for charge compensation. Analyses were performed in automated sequences, with each analysis comprising a 70 second pre-sputter to remove the gold coating over a rastered 15×15 µm area, centering of the secondary beam in the field aperture to correct for small variations in surface relief, and data acquisition in 16 four-second integration cycles. The magnetic field was locked at the beginning of the session using a nuclear magnetic resonance (NMR) field sensor. Secondary ion signals for ^{32}S and ^{34}S were detected simultaneously using two Faraday detectors with a common mass resolution of 4860 (M/ΔM). Data were normalized for instrumental mass fractionation using matrix matched reference materials which were mounted together with the sample mounts and analyzed after every sixth sample analysis. Results are reported as per mil (‰) $\delta^{34}S$ based on the V-CDT reference value [50]. Analytical transects of up to ten spots were made from core to rim in the crystals. In total, 89 analyses were made for $\delta^{34}S$ of pyrite from 11 crystals from three fracture samples. The pyrite reference material S0302A with a conventionally determined value of 0.0‰ ± 0.2‰ (R. Stern, University of Alberta, pers. comm.) was used. Typical precision on a single $\delta^{34}S$ value, after propagating the within run and external uncertainties from the reference material measurements was ±0.07‰.

For calcite, a total number of 84 $\delta^{13}C$ and 93 for $\delta^{18}O$ SIMS-analyses were performed on the same Cameca IMS1280 described above. Settings follow those described for S isotopes, with some differences: O was measured on two Faraday cups (FC) at mass resolution 2500, C used a FC/Electron Multiplier combination with mass resolution 2500 on the 12C peak and 4000 on the ^{13}C peak to resolve it from $^{12}C^{1}H$. Calcite results are reported as per mil (‰) $\delta^{13}C$ based on the Pee Dee Belemnite (V-PDB) reference value. Analyses were carried out running blocks of six unknowns bracketed by two standards. Analytical transects of up to nine spots were made from core to rim in the crystals. Up to five crystals were analyzed from each fracture sample. Analyses were made for 31 crystals from 8 fracture samples. Isotope data from calcite were normalized using calcite reference material S0161 from a granulite facies marble in the Adirondack Mountains, kindly provided by R.A. Stern (Univ. of Alberta). The values used for instrumental mass fractionation correction were determined by conventional stable isotope mass spectrometry at Stockholm University on ten separate pieces, yielding $\delta^{13}C = 0.22‰ ± 0.35‰$ V-PDB (1 std. dev.) and $\delta^{18}O = −5.62‰ ± 0.22‰$ V-PDB (1 std. dev.). Precision was $\delta^{18}O$: ± 0.2‰ − 0.3‰ and $\delta^{13}C$: ± 0.4‰–0.5‰. Values of the reference material measurements are listed together with the samples in Supplementary Table S1 ($\delta^{13}C$), Table S2 ($\delta^{18}O$), and Table S3 ($\delta^{34}S$).

3.3. Laser Ablataion Inductively Coupled Plasma Mass Spectrometry (LA-ICP-MS) for U-Pb Geochronology

U-Pb geochronology via the in situ LA-ICP-MS method was conducted at the Geochronology and Tracers Facility, British Geological Survey (Nottingham, UK). The method utilizes a New Wave Research 193UC excimer laser ablation system, coupled to a Nu Instruments Attom single-collector sector-field ICP-MS. The method follows that previously described in [51], and involves a standard-sample bracketing with normalization to NIST 614 silicate glass for Pb-Pb ratios and WC-1 carbonate for U-Pb ratios. The laser parameters comprise a 80 μm static spot, fired at 10 Hz, with a ~6 J/cm^2 fluence, for 20 s of ablation. Material is pre-ablated to clean the sample site with 120 μm spots for 2 s. No common lead correction is made; ages are determined by regression and the lower intercept on a Tera-Wasserburg plot (using Isoplot 4.15; [52]). Duff Brown, a carbonate previously measured by isotope dilution mass spectrometry was used as a validation, and pooling of all sessions yields a lower intercept age of 64.2 ± 1.6 Ma (MSWD = 4.0), overlapping the published age of 64.04 ± 0.67 Ma [53]. All ages are plotted and quoted at 2σ and include propagation of systematic uncertainties according to the protocol described in Horstwood et al., [54]. Data are screened for low Pb and low U counts below detection, and very large uncertainties on the Pb-Pb and Pb-U ratios which indicate mixed analyses. The spots are also checked after ablation for consistent ablation pit shape, and data are rejected if the ablations were anomalous (this results from material cleaving off, or clipping the resin mount).

Eight samples of calcite were screened from the COSC-1 drill core, but only one sample yielded measurable radiogenic lead. This sample (178 m) yielded variably robust U-Pb ages in the first session (based on the SIMS-analyzed crystals), and so a second session was added to this dataset using further mounted crystals. Full analytical data from the sessions are listed in Supplementary Table S4).

4. Results

4.1. Mineralogy

Mineralogical composition was characterized in eight samples, three from micro-karst (122, 122.8, 178 m) and five from deeper veins (684, 743, 1051, 1369, 2210 m). The micro-karst samples were investigated in the SEM in the karst cavities of the porous rocks in cut-off rock chips. The deeper veins were studied in the SEM as polished blocks (Figure 2).

4.1.1. Micro-Karst

The micro-karsts occurred as foliation parallel cavities in the gneiss (Figure 2a,c, Supplementary Table S5) and contained secondary fine-grained euhedral (rhombohedral) calcite crystals of up to 400 μm in size (Figure 2b,d) that line the walls of the micro-karst cavities. The calcite crystals commonly occur in aggregates of several equant crystals (see also Figure S2a–d). No growth zonations or overgrowths are evident based on the observation of morphology of the crystals in situ within the voids. Subhedral to euhedral pyrite crystals of up to 400 μm size occur together with calcite in the karst voids (Figure 2b,d). In sample 178 m, the zeolite chabazite is abundant, as euhedral blocky crystals of up to 1 mm in size (Figure 2d, Figure S2b–d). Chabazite is commonly intergrown with calcite. Albite and radiating aggregates of platy clay mineral crystals (Figure S2d) are occasionally present, and very fine-grained Ni-rich grains, which have detectable amounts of C, P and Si (EDS-analysis). Fine-grained C-rich filaments occur, particularly in sample 122.8, but also in sample 178 m (see SEM-images in Supplementary Figure S3). The polished calcite cross sections from the micro-karst revealed only scattered single-phased inclusions.

4.1.2. Veins

The sampled calcite veins were 1–10 mm in width (Figure 2e,g, more sample photos are shown in Supplementary Table S6), and occur both as individual veins and as breccias with calcite cement and wall rock fragments. Apart from calcite, the veins contain quartz, albite, chlorite and sulfides (pyrite and/or pyrrhotite) (Figure 2e–h). In sample 1051, albite lines the fracture and calcite is in the vein center. In sample 684, dolomite is abundant together with calcite (Figure 2f). Talc, pyrite and Ti-oxide (Figure S2e) also occur in this sample intergrown with the carbonates, suggesting formation of this mineral assemblage at a single event. There is also a younger calcite-talc filled micro-vein running along the edge of the main vein of this sample. Ca-Al-silicates occur in two samples, as epidote in sample from 1369 m occurring together with quartz (lining the fracture), calcite and pyrrhotite (Figure S2g), and as laumontite in sample from 743 m, where it is intergrown with calcite (Figure S2f). The deepest sample (2210 m) holds secondary rare earth element (REE)-carbonate and ilmenite, intergrown with albite (Figure S2h). See Supplementary Tables S5 and S6 for detailed sample descriptions.

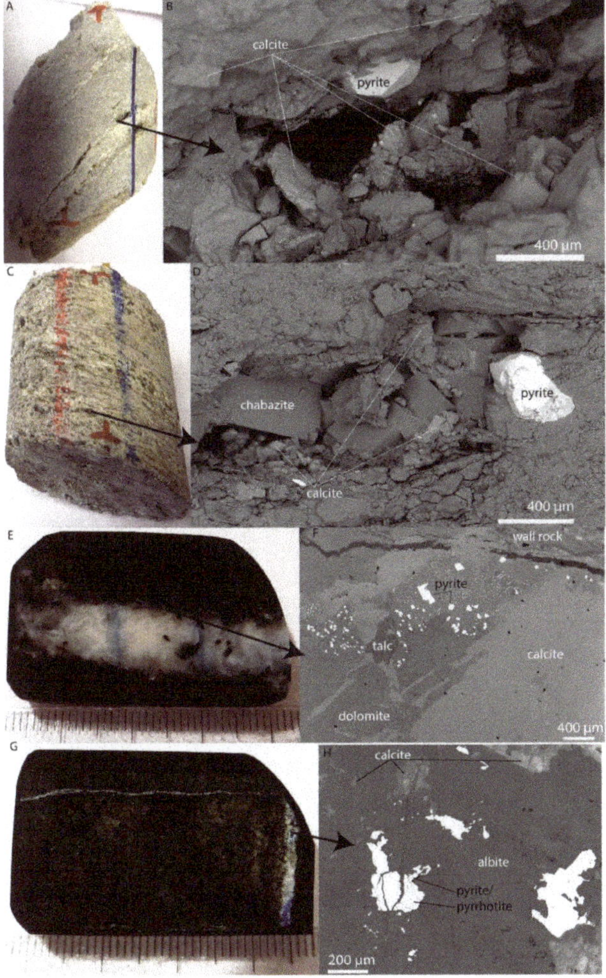

Figure 2. Mineral characteristics in photographed drill core samples (left) and back-scattered scanning electron microscope (SEM) images (right). (**A–B**): sample 122, micro-karst, showing secondary calcite and pyrite in a cavity. (**C–D**): sample 178, micro-karst, showing secondary chabazite, calcite and pyrite in a cavity. (**E–F**) calcite vein that also holds dolomite, talc and pyrite, sample 684. (**G–H**) vein with abundant albite, calcite and sulfides (intergrown pyrite and pyrrhotite), sample 1051.

4.2. Stable Isotopes

Calcite shows stable isotope compositions of δ^{18}O: −28.9‰ to −10.7‰ V-PDB (span 18.2‰, n_{spots} = 93, $n_{samples}$ = 8) and δ^{13}C: −17.4‰ to −3.5‰ V-PDB (span 13.9‰, n_{spots} = 84, $n_{samples}$ = 8). The karst calcite generally shows different populations of C and O isotope composition than the calcite veins (Figure 3a). The veins generally have depleted δ^{18}O values, mainly lighter than −22‰ (Figure 3b), and δ^{13}C values that are quite invariable at −8‰ ± 1‰ V-PDB in the two deepest samples and a larger variability in the three shallowest veins (−17‰ to −5‰ V-PDB Figure 3c). The shallow veins also show larger spans in $\delta^{18}O$ than the deeper two (Figure 3b).

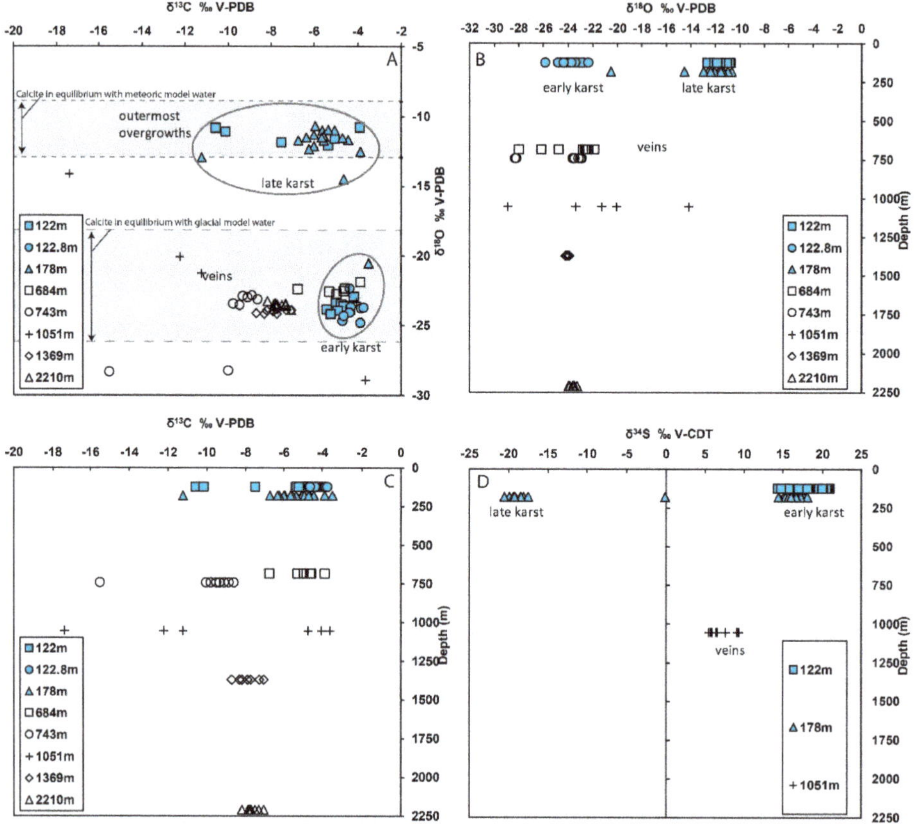

Figure 3. Stable isotope scatter and depth plots, divided into micro-karst samples (blue symbols) and vein samples (transparent symbols and black crosses). (**A**) $\delta^{18}O_{calcite}$ vs. $\delta^{18}O_{calcite}$, with marker for the different micro-karst groups (spheres), as well as spans of calculated values for hypothetical calcite precipitated from ambient meteoric or glacial water. For this comparison, temperature dependent fractionation factors for calcite precipitation at ambient temperatures are used for $\delta^{18}O$ [22]. See text for details of temperatures and isotope spans of the model waters. (**B**) $\delta^{18}O_{calcite}$ vs. depth. (**C**) $\delta^{13}C_{calcite}$ vs. depth. (**D**) $\delta^{34}S_{pyrite}$ vs. depth.

Two main groups of micro-karst calcite are evident, particularly manifested by the $\delta^{18}O$ composition, which shows a cluster at −25‰ to −20‰ V-PDB with most values below −22.5‰, and one cluster with heavier values, at −14.5‰ to −10.8‰ V-PDB, with most values between −13‰ and −11‰. The latter group occurs in samples 122 and 178, but not in sample 122.8 (Figure 3a). In sample 178, the isotopically heavy $\delta^{18}O$ values dominate. There is petrographic evidence for a temporal trend in the $\delta^{18}O$ values. The isotopically light $\delta^{18}O$ group is older than the isotopically heavier group as shown in Figure 4 where the core of the crystals is ^{18}O-poor compared to the later growth zone(s). In sample 122, the relatively ^{18}O-rich overgrowths make up very small parts of the crystal volume (Figure 4a–d), in contrast to in sample 178 where most of the crystal volume is made up of the ^{18}O-rich younger calcite (Figure 4e,f). The outermost overgrowth in sample 122 (with high Back Scattered Electron-intensity, Figure 4a,c) features a significant decrease in ^{13}C compared to the other parts (Figure 4b,d). This means that three generations are observed within the calcite in the micro-karst (1: early karst, 2: late karst, 3: outermost overgrowth of late karst calcite, Figure 3a). A similar overgrowth occurs in sample 178 (δ^{13}C: −11.2‰), but only in one spot, showing that the second calcite group dominated this sample.

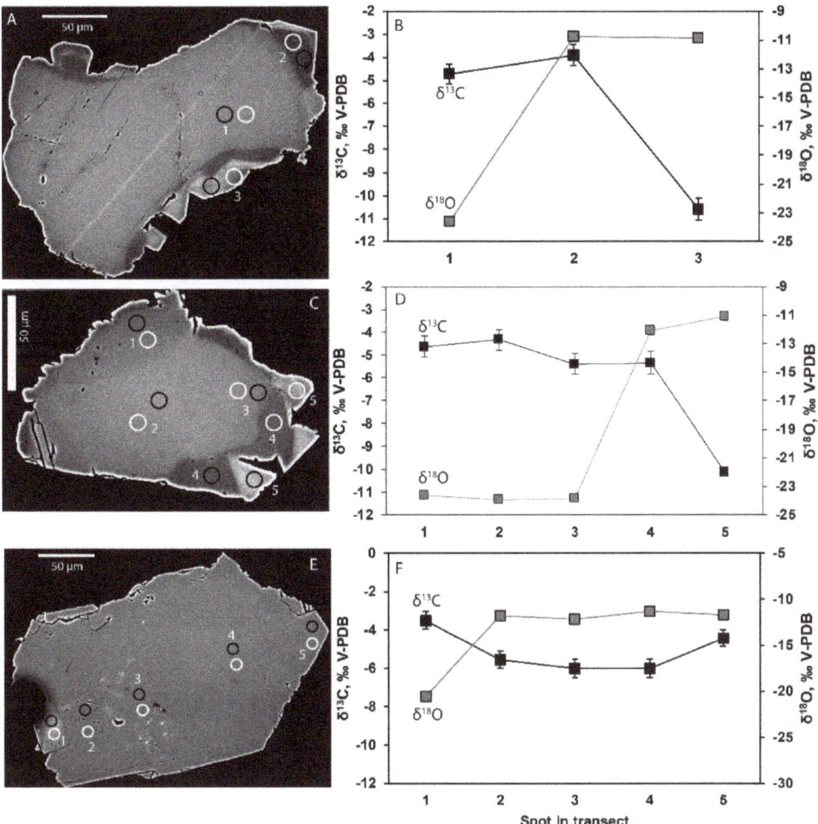

Figure 4. Microanalytical secondary ion mass spectrometry (SIMS) transects within calcite. Back scattered electron images of polished crystal cross sections are shown to the left (**A,C,E**) with spot locations indicated, for closely spaced 10 μm δ^{13}C (black) and δ^{18}O (white) SIMS spots. Corresponding isotopic values are shown on the graphs in **B** (sample 122 m), **D** (122 m), and **F** (178 m).

Pyrite shows δ^{34}S compositions ranging from −20.5‰ to +20.9‰ V-CDT (span 41.4‰, n_{spots} = 89, $n_{samples}$ = 3). Two samples show very narrow span, vein 1051 m: 7.8‰ ± 1.6‰ and micro-karst 122 m: 18.8‰ ± 2.2‰. The karst sample 178 m shows two distinct groups (−19.8‰ ± 1.6‰ and +16.4‰ ± 1.0‰) and one value in between (−0.1‰), of which the isotopically heaviest group overlaps with micro-karst sample 122 m. Petrographically, the isotopically heavy pyrite in sample 178 m is oldest, as it is found in the inner part of a crystal with an isotopically light rim (Figure 5a). The same sample also features pyrite crystals having both light values throughout the whole crystal (Figure 5B) as well as heavy values throughout the whole crystal (Table S3). The isotopically light pyrite is coeval with the calcite zonation group "late karst" with δ^{13}C values of −7‰ to −3‰.

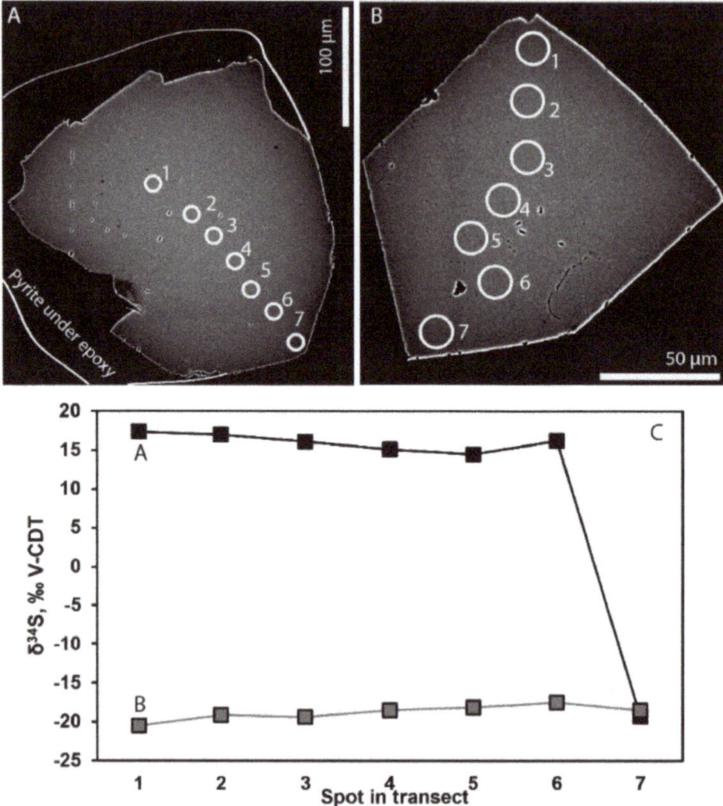

Figure 5. Microanalytical SIMS δ^{34}S transects within pyrite. Back Scattered Electron images of polished crystal cross sections are shown in **A** and **B** with spot locations indicated, for closely spaced 10 μm δ^{34}S SIMS spots. Corresponding isotopic values are shown in **C**. In A, a part of the crystal is not exposed in the cross-section, as indicated by the line.

4.3. U-Pb Geochronology

The only calcite sample yielding a robust U-Pb age determination was 178 m. The spatial correlation between the SIMS spots for stable isotopes and the U-Pb LA-ICP-MS spots shows that the dating represents the late karst calcite group in Figure 3a (most δ^{18}O values between −13‰ and −11‰ and δ^{13}C of −7‰ to −4‰), but not the outermost overgrowths. The LA-ICP-MS spots from a couple of different crystals within the late karst calcite population line up along two distinct regressions (Figure 6), with ages of 9.6 ± 1.3 Ma (MSWD = 1.4) and 2.5 ± 0.2 Ma (MSWD = 0.9). Each of these regressions represents several spots from multiple crystals, but each crystal only conforms to one of the ages. The latest calcite overgrowth and the older crystal cores were not targeted in this sample due to the small size of the crystal domains of these groups.

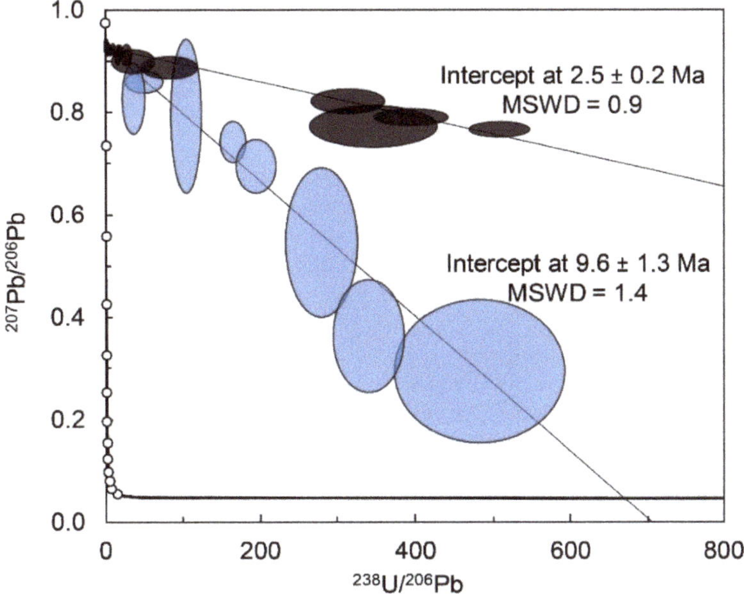

Figure 6. U-Pb dating of calcite sample 178 m, ages reflect non-anchored lower intercept ages and uncertainties are quoted at 2 s and comprise systematic uncertainties. Spot locations and details are reported in Supplementary Table S4 and Figure S4.

5. Discussion

The different populations of isotope compositions in the mineral veins and micro-karst precipitates suggests several discrete events of precipitation and fluid circulation in the fracture system of the thrust sheets.

5.1. Microbial Activity

Microbial activity typically results in distinguishable C isotopic excursions [15] that can be traced in authigenic carbonate minerals [10,55]. The samples in this study show very moderate C-isotope variations, especially compared to fracture coatings from other sites on the Fennoscandian shield [26,28,56]. The C isotope signatures in the calcites of the COSC-1 core are thus no strong marker for microbial activity in situ, at least not for processes involving methane formation or oxidation, which are known to produce carbonates strongly enriched in ^{13}C or depleted in ^{13}C, respectively [8,15,55,57], as documented elsewhere on the Fennoscandian shield [26,28,29,58,59]. The moderate δ^{13}C-depletion that is observed may, however, reflect the influence of C originating from dissolved organic carbon that has been oxidized by microbial communities and then mixed with less ^{13}C-depleted dissolved inorganic carbon in the waters. The C isotope composition of calcite reflects the result of several (cryptic) potential processes, and interpretations of δ^{13}C compositions in the observed span are not straightforward without additional evidence. Remnants of a biofilm occur on the karst cavity walls in sample 122 (Figure S3) and is another indication of microbial activity. The SEM-EDS analysis confirmed a carbonaceous composition, but the sample volume of the biofilm was too small for biomarker analysis, which could have offered further information about what communities inhabited this cavity. The small variation and small depletion in δ^{13}C values of the two deepest samples, suggest very small, if any, influence from microbial activity here and formation from a single precipitation event and fluid source, in contrast to the shallower samples (Figure 3c).

The generally large fractionation of the stable sulfur isotopes associated with microbial sulfate reduction, due to faster turnover of ^{32}S than ^{34}S [60], thus leading to discrimination of ^{34}S, has made this isotope system one of the most extensively used for understanding both modern and ancient biogeochemical cycles [61,62]. Laboratory culture measurements have reported sulfur isotope enrichments ($\delta^{34}S_{\text{sulfate-sulfide}}$ or $^{34}\varepsilon$) as large as 66‰ [63] but even larger have been inferred from natural observations [64–66].

The low minimum $\delta^{34}S_{\text{pyrite}}$ values ($-19.8‰ \pm 1.6‰$) in the young pyrite population reflect ^{32}S enrichment in the produced sulfide during MSR in the micro-karst. The initial sulfate δ^{34}S composition is unknown, but ambient water within the Fennoscandian Shield at other sites have $\delta^{34}S_{SO4}$ values in the +15‰ to +25‰ range [67], which implies an isotope enrichment, $^{34}\varepsilon$ ($\delta^{34}S_{SO4}$-$\delta^{34}S_{\text{pyrite}}$), of 35–45‰ if we assumed pyrite in the studied fractures formed from water of similar $\delta^{34}S_{SO4}$ composition. This $^{34}\varepsilon$ is fully in line with MSR [66]. This younger generation of pyrite is related to the second and dominant calcite in the micro-karst at 178 m. A calcite generation that gave two different ages for two grain populations (2.5 ± 0.2 Ma and 9.6 ± 1.3 Ma) which thus also represents the age span of this MSR-related pyrite generation. The relatively narrow span in values for this pyrite suggests that it formed under open system conditions because there is no indication of successively increasing values with growth, which is a typical feature of Rayleigh isotope fractionation at closed or semi-closed conditions for sulfate [12].

The older pyrite generation in the micro-karst (dominating in sample at 122 m depth) showing more ^{34}S-enriched values can be thermochemical in origin, as this process generally does not involve significant degrees of fractionation [16,68]. Alternatively, the δ^{34}S values of this generation can reflect a late stage MSR-system undergoing Rayleigh fractionation in a semi-closed system, but this requires that significant amounts of isotopically light sulfide has been produced and precipitated elsewhere along the flow path [14].

5.2. Paleo-Fluid Flow and Water Types

The O isotope composition of the calcite crystals and veins can provide information about the fluid source. Unfortunately, there is no hydrochemical data available from the deep fractures in the COSC-1 borehole for comparison. Instead, we compare our determined δ^{18}O values with model glacial and meteoric waters at ambient temperatures, taking into account the temperature-dependent O-isotope fractionation that occurs during calcite precipitation (Figure 3a, [22]. For the late micro-karst precipitates, there is no temperature estimate available due to the fact that no two-phased fluid inclusions could be detected. This feature points to fluid inclusion entrapment at temperatures below 50 °C [69], and the detection of coeval MSR-related $\delta^{34}S_{\text{pyrite}}$ values also is in favor of low-temperature formation. If we assign a hypothetical meteoric water (δ^{18}O: $-11.5‰$ to $-9.5‰$ SMOW) based on modern precipitation in Sweden [67] and formation temperatures of 5 to 20 °C, it is in accordance with the composition of the $2.5 \pm 0.2 / 9.6 \pm 1.3$ Ma micro-karst calcite (Figure 3a). This means that precipitation of this calcite from a meteoric water is possible. However, it should be noted that in high latitude terrains, the δ^{18}O values can be even lighter than those we assigned [70]. This period of fluid flow and mineral growth in the micro-karst is temporally related to late Oligocene to Pliocene continental uplift of Fennoscandia [71] and the present relief of the Caledonides is largely the result of uplift during these times [35] with subsequent glacial-erosion modification [34].

The δ^{18}O composition of glacial meltwater from Pleistocene ice sheets in Fennoscandia has been estimated to a range of $-22‰$ to $-20‰$ [72], but recent studies of sub-ice sheet runoff at western Greenland show even more depleted values. When assigning a hypothetical glacial water with δ^{18}O of $-27‰$ to $-20‰$ SMOW, and temperatures of 5–10 °C, the temperature dependent fractionation results in modeled values in calcite of c. $-26‰$ to $-18‰$ V-PDB (Figure 3a), which is significantly lighter than the young micro-karst calcite, but in line with the early micro-karst and vein calcite. This means that it is unlikely that the young micro-karst calcite precipitated from a glacial water, but that the older calcite and veins may be glacial precipitates. However, if we assign a brine with similar composition as at

Sellafield (δ^{18}O: -5‰ SMOW [73], Figure 1) and higher formation temperatures (100–150 °C), we will end up with hypothetical calcite with δ^{18}O values (c. -23 to -19 V-PDB) that also match the most ^{18}O-rich precipitates of these calcite groups. Furthermore, the mineralogical assemblage of the veins, with minerals such as epidote and laumontite, that are not formed at ambient temperatures [74,75], also speak against low-temperature formation from glacial water. For future studies, we suggest that the clumped isotope methodology is utilized for COSC-veins in order to potentially obtain formation temperature estimates of the carbonates. This approach has recently been proven successful to determine calcite precipitation temperatures in veins and constrain fluid sources, and fluxes, e.g., in the Peak District, UK [76]. Taken together, the calcites can be divided into groups based on their isotopic composition (mainly decided on δ^{18}O) of which the veins seems to be of higher temperature type based on the presence of epidote and laumontite. The young micro-karst calcite was the only group that was dated ($2.5 \pm 0.2 / 9.6 \pm 1.3$ Ma) and overlaps with a low-temperature meteoric water. There is also an even younger group of overgrowths (i.e., younger than 2.5 ± 0.2 Ma) with lower δ^{13}C values than the dated calcite, but the overgrowths were too small for U-Pb LA-ICP-MS dating.

6. Conclusions

We present a microanalysis study (SIMS and LA-ICP-MS) recording ancient secondary mineral formation, fluid flow and microbial activity in a nappe unit of the Scandinavian Caledonides (utilizing the deep COSC-1 borehole). Petrographic and isotopic evidence for several generations of calcite and pyrite growth in micro-karst are determined. ^{34}S-depleted composition of pyrite suggests formation following microbial sulfate reduction in the karst and the O isotope composition of coeval calcite dated to $2.5 \pm 0.2 / 9.6 \pm 1.3$ Ma overlap with low-temperature meteoric water, in contrast to older calcite that has significantly lower δ^{18}O values. Our results mark the potential of combined SIMS and LA-ICP-MS micro-analysis within fine-grained mineral grains to understand and temporally determine discrete events of fluid flow and microbial activity in deep fracture systems and karsts of mountain ranges.

Supplementary Materials: The following are available online at http://www.mdpi.com/2076-3263/10/2/56/s1: Table S1: SIMS analyses of C isotopes in calcite and matrix matched reference material; Table S2: SIMS analyses of O isotopes in calcite and matrix matched reference material; Table S3: SIMS analyses of S isotopes in pyrite and matrix matched reference material; Table S4: LA-ICP-MS analyses for U-Pb calcite geochronology of sample 178 and matrix matched reference materials. Table S5: Sample photo documentation. Table S6: Sample details. Supplementary Figures S1–S4 with captions; Figure S1. Core log with mineral samples marked. This log is based on on-site descriptions, and the depth is subject to minor corrections post-drilling (this file is adopted from the 'operation datasets' for COSC-1, http://dataservices.gfz-potsdam.de/icdp/showshort.php?id=escidoc:1095929). Figure S2. Back-scattered SEM-images showing (A) aggregates of calcite crystals on the walls of micro-karst cavities of sample 122 m depth, (B) Euhedral and partly intergrown crystals of chabazite and calcite on the walls of micro-karst cavities of sample 178 m depth. Chabazite is slightly darker and not as fine-grained as calcite. (C) Euhedral and partly intergrown crystals of chabazite and calcite on the walls of micro-karst cavities of sample 178 m depth. (D) Euhedral and partly intergrown crystals of clay minerals (radiating aggregate of platy crystals), chabazite and calcite on the walls of micro-karst cavities of sample 178 m depth. (E) Vein assemblage of calcite, chlorite and Ti-oxide in polished block of sample 684 m. (F) Vein assemblage of calcite and laumontite (darker grey than calcite) in polished block of sample 743 m. Note that laser-ablation spots (visible) for dating tries have altered the sample slightly (darker areas). (G) Vein assemblage of calcite, pyrrhotite, quartz and epidote in polished block of sample 1369 m. (H) Vein assemblage of calcite, albite and ilmenite in polished block of sample 2210 m. Figure S3. Back-scattered SEM-images showing remnants of biofilm (carbonaceous matter) on the walls of micro-karst cavities of sample 122.8 m depth. Figure S4. Spot locations for LA-ICP-MS U-Pb geochronology analyses in polished calcite grains of sample 178 m, numbers and colors correspond to analytical details in Table S4.

Author Contributions: Conceptualization, H.D. and N.M.W.R.; methodology, and formal analysis, H.D. N.M.W.R. and M.J.W; investigation, H.D. and N.M.W.R. writing—original draft preparation, H.D. and N.M.W.R.; writing—review and editing, H.D., N.M.W.R. and M.J.W; visualization, H.D. and N.M.W.R.; funding acquisition, H.D., N.M.W.R. and M.J.W. All authors have read and agreed to the published version of the manuscript.

Funding: This research was funded by Swedish research council (contract 2017-05186 to H.D.) and Formas (contract 2017-00766 to H.D. and M.J.W.). The Natural Environment Research Council provides financial support of the Geochronology and Tracers Facility, British Geological Survey. NordSIM is part of the NordSIM-Vegacenter infrastructure supported by the Swedish Research Council grant 2017-00637. COSC-1 was drilled within the Swedish national research infrastructure for scientific drilling "Riksriggen" and financed by the International Continental Scientific Drilling Program (ICDP) and the Swedish Research Council.

Acknowledgments: K. Lindén and H. Jeon are thanked for analytical and/or sample preparation assistance. M. Konrad-Schmolke at University of Gothenburg is thanked for SEM analysis access. H. Lorenz provided visualization support. N.M.W.R. thanks the British Geological Survey for funding to sample the core, and John Fletcher for sample preparation. This is NordSIM publication 630.

Conflicts of Interest: The authors declare no conflict of interest.

References

1. De Graaf, S.; Lüders, V.; Banks, D.A.; Sośnicka, M.; Reijmer, J.J.; Kaden, H.; Vonhof, H.B. Fluid evolution and ore deposition in the Harz Mountains revisited: Isotope and crush-leach analyses of fluid inclusions. *Miner. Depos.* **2019**, 1–16. [CrossRef]

2. Ivarsson, M.; Kilias, S.P.; Broman, C.; Neubeck, A.; Drake, H.; Fru, E.C.; Bengtson, S.; Naden, J.; Detsi, K.; Whitehouse, M.J. Exceptional Preservation of Fungi as H2-Bearing Fluid Inclusions in an Early Quaternary Paleo-Hydrothermal System at Cape Vani, Milos, Greece. *Minerals* **2019**, *9*, 749. [CrossRef]

3. Dexter-Dyer, B.; Kretzschmar, M.; Krumbein, W.E. Possible microbial pathways in the formation of Precambrian ore deposits. *J. Geol. Soc.* **1984**, *141*, 251–262. [CrossRef]

4. Martini, A.M.; Budai, J.M.; Walter, L.M.; Schoell, M. Microbial generation of economic accumulations of methane within a shallow organic-rich shale. *Nature* **1996**, *383*, 155–158. [CrossRef]

5. Drake, H.; Roberts, N.M.W.; Heim, C.; Whitehouse, M.J.; Siljeström, S.; Kooijman, E.; Broman, C.; Ivarsson, M.; Åström, M.E. Timing and origin of natural gas accumulation in the Siljan impact structure, Sweden. *Nat. Commun.* **2019**, *10*, 4736. [CrossRef] [PubMed]

6. Tullborg, E.-L.; Drake, H.; Sandström, B. Palaeohydrogeology: A methodology based on fracture mineral studies. *Appl. Geochem.* **2008**, *23*, 1881–1897. [CrossRef]

7. Sahlstedt, E.; Karhu, J.A.; Pitkänen, P. Indications for the past redox environments in deep groundwaters from the isotopic composition of carbon and oxygen in fracture calcite, Olkiluoto, SW Finland. *Isot. Environ. Heal. Stud.* **2010**, *46*, 370–391. [CrossRef] [PubMed]

8. Budai, J.M.; Walter, L.M.; Ku, T.C.W.; Martini, A.M. Fracture?fill calcite as a record of microbial methanogenesis and fluid migration: A case study from the Devonian Antrim Shale, Michigan Basin. *Geofluids* **2002**, *2*, 163–183. [CrossRef]

9. Feng, D.; Chen, D.; Peckmann, J.; Bohrmann, G. Authigenic carbonates from methane seeps of the northern Congo fan: Microbial formation mechanism. *Mar. Pet. Geol.* **2010**, *27*, 748–756. [CrossRef]

10. Natalicchio, M.; Birgel, D.; Pierre, F.D.; Martire, L.; Clari, P.; Spötl, C.; Peckmann, J. Polyphasic carbonate precipitation in the shallow subsurface: Insights from microbially-formed authigenic carbonate beds in upper Miocene sediments of the Tertiary Piedmont Basin (NW Italy). *Palaeogeogr. Palaeoclim. Palaeoecol.* **2012**, *329*, 158–172. [CrossRef]

11. Caesar, K.H.; Kyle, J.R.; Lyons, T.W.; Tripati, A.; Loyd, S.J. Carbonate formation in salt dome cap rocks by microbial anaerobic oxidation of methane. *Nat. Commun.* **2019**, *10*, 808. [CrossRef] [PubMed]

12. Kohn, M.J.; Riciputi, L.R.; Stakes, D.; Orange, D.L. Sulfur isotope variability in biogenic pyrite; reflections of heterogeneous bacterial colonization? *Am. Mineral.* **1998**, *83*, 1454–1468. [CrossRef]

13. Parnell, J.; Boyce, A.; Thackrey, S.; Muirhead, D.; Lindgren, P.; Mason, C.; Taylor, C.; Still, J.; Bowden, S.; Osinski, G.R.; et al. Sulfur isotope signatures for rapid colonization of an impact crater by thermophilic microbes. *Geology* **2010**, *38*, 271–274. [CrossRef]

14. Drake, H.; Whitehouse, M.J.; Heim, C.; Reiners, P.W.; Tillberg, M.; Hogmalm, K.J.; Dopson, M.; Broman, C.; Åström, M.E. Unprecedented 34 S-enrichment of pyrite formed following microbial sulfate reduction in fractured crystalline rocks. *Geobiology* **2018**, *16*, 556–574. [CrossRef]

15. Knittel, K.; Boetius, A. Anaerobic Oxidation of Methane: Progress with an Unknown Process. *Annu. Rev. Microbiol.* **2009**, *63*, 311–334. [CrossRef]

16. Machel, H.G.; Krouse, H.R.; Sassen, R. Products and distinguishing criteria of bacterial and thermochemical sulfate reduction. *Appl. Geochem.* **1995**, *10*, 373–389. [CrossRef]

17. Drake, H.; Tullborg, E.-L. Paleohydrogeological events recorded by stable isotopes, fluid inclusions and trace elements in fracture minerals in crystalline rock, Simpevarp area, SE Sweden. *Appl. Geochem.* **2009**, *24*, 715–732. [CrossRef]

18. Blyth, A.; Frape, S.; Ruskeeniemi, T.; Blomqvist, R. Origins, closed system formation and preservation of calcites in glaciated crystalline bedrock: Evidence from the Palmottu natural analogue site, Finland. *Appl. Geochem.* **2004**, *19*, 675–686. [CrossRef]

19. Mathurin, F.A.; Åström, M.E.; Laaksoharju, M.; Kalinowski, B.E.; Tullborg, E.-L. Effect of Tunnel Excavation on Source and Mixing of Groundwater in a Coastal Granitoidic Fracture Network. *Environ. Sci. Technol.* **2012**, *46*, 12779–12786. [CrossRef]

20. O'Neil, J.R. Oxygen Isotope Fractionation in Divalent Metal Carbonates. *J. Chem. Phys.* **1969**, *51*, 5547. [CrossRef]

21. Watkins, J.M.; Hunt, J.D.; Ryerson, F.J.; DePaolo, D.J. The influence of temperature, pH, and growth rate on the δ18O composition of inorganically precipitated calcite. *Earth Planet. Sci. Lett.* **2014**, *404*, 332–343. [CrossRef]

22. Kim, S.-T.; O'Neil, J.R. Equilibrium and nonequilibrium oxygen isotope effects in synthetic carbonates. *Geochim. Cosmochim. Acta* **1997**, *61*, 3461–3475. [CrossRef]

23. Drake, H.; Ivarsson, M.; Tillberg, M.; Whitehouse, M.J.; Kooijman, E. Ancient Microbial Activity in Deep Hydraulically Conductive Fracture Zones within the Forsmark Target Area for Geological Nuclear Waste Disposal, Sweden. *Geosciences* **2018**, *8*, 211. [CrossRef]

24. Drake, H.; Ivarsson, M.; Bengtson, S.; Heim, C.; Siljeström, S.; Whitehouse, M.J.; Broman, C.; Belivanova, V.; Åström, M.E. Anaerobic consortia of fungi and sulfate reducing bacteria in deep granite fractures. *Nat. Commun.* **2017**, *8*, 55. [CrossRef]

25. Drake, H.; Åström, M.E.; Tullborg, E.-L.; Whitehouse, M.; Fallick, A.E. Variability of sulphur isotope ratios in pyrite and dissolved sulphate in granitoid fractures down to 1km depth—Evidence for widespread activity of sulphur reducing bacteria. *Geochim. Cosmochim. Acta* **2013**, *102*, 143–161. [CrossRef]

26. Sahlstedt, E.; Karhu, J.A.; Pitkänen, P.; Whitehouse, M. Biogenic processes in crystalline bedrock fractures indicated by carbon isotope signatures of secondary calcite. *Appl. Geochem.* **2016**, *67*, 30–41. [CrossRef]

27. Sahlstedt, E.; Karhu, J.; Pitkanen, P.; Whitehouse, M. Implications of sulfur isotope fractionation in fracture-filling sulfides in crystalline bedrock, Olkiluoto, Finland. *Appl. Geochem.* **2013**, *32*, 52–69. [CrossRef]

28. Drake, H.; Heim, C.; Roberts, N.M.; Zack, T.; Tillberg, M.; Broman, C.; Ivarsson, M.; Whitehouse, M.J.; Åström, M.E. Isotopic evidence for microbial production and consumption of methane in the upper continental crust throughout the Phanerozoic eon. *Earth Planet. Sci. Lett.* **2017**, *470*, 108–118. [CrossRef]

29. Drake, H.; Åström, M.E.; Heim, C.; Broman, C.; Åström, J.; Whitehouse, M.; Ivarsson, M.; Siljestrom, S.; Sjövall, P. Extreme 13C depletion of carbonates formed during oxidation of biogenic methane in fractured granite. *Nat. Commun.* **2015**, *6*, 7020. [CrossRef]

30. Lorenz, H.; Rosberg, J.-E.; Juhlin, C.; Bjelm, L.; Almqvist, B.S.G.; Berthet, T.; Conze, R.; Gee, D.G.; Klonowska, I.; Pascal, C.; et al. COSC-1 – drilling of a subduction-related allochthon in the Palaeozoic Caledonide orogen of Scandinavia. *Sci. Drill.* **2015**, *19*, 1–11. [CrossRef]

31. Gee, D.G.; Juhlin, C.; Pascal, C.; Robinson, P. Collisional Orogeny in the Scandinavian Caledonides (COSC). *GFF* **2010**, *132*, 29–44. [CrossRef]

32. Corfu, F.; Andersen, T.B.; Gasser, D. The Scandinavian Caledonides: Main features, conceptual advances and critical questions. *Geol. Soc. Lond. Spéc. Publ.* **2014**, *390*, 9–43. [CrossRef]

33. Roberts, D. The Scandinavian Caledonides: Event chronology, palaeogeographic settings and likely modern analogues. *Tectonophysics* **2003**, *365*, 283–299. [CrossRef]

34. Lidmar-Bergström, K.; Näslund, J.O.; Ebert, K.; Neubeck, T.; Bonow, J.M. Cenozoic landscape development on the passsive margin of northern Scandinavia. *Nor. J. Geol.* **2007**, *87*, 181–196.

35. Lidmar-Bergström, K. Long term morphotectonic evolution in Sweden. *Geomorphology* **1996**, *16*, 33–59. [CrossRef]

36. Kendrick, M.; Burgess, R.; Harrison, D.; Bjørlykke, A. Noble gas and halogen evidence for the origin of Scandinavian sandstone-hosted Pb-Zn deposits. *Geochim. Cosmochim. Acta* **2005**, *69*, 109–129. [CrossRef]

37. Hedin, P.; Almqvist, B.; Berthet, T.; Juhlin, C.; Buske, S.; Simon, H.; Giese, R.; Krauß, F.; Rosberg, J.-E.; Alm, P.-G. 3D reflection seismic imaging at the 2.5km deep COSC-1 scientific borehole, central Scandinavian Caledonides. *Tectonophysics* **2016**, *689*, 40–55. [CrossRef]

38. Lorenz, H.; Rosberg, J.E.; Juhlin, C.; Bjelm, L.; Almqvist, B.G.S.; Berthet, T.; Conze, R.; Gee, D.; Klonowska, I.; Pascal, C.; et al. Operational report about phase 1 of the Collisional Orogeny in the Scandinavian Caledonides scientific drilling project (COSC-1). *ICDP Oper. Rep. Potsdam GFZ Ger. Res. Cent. Geosci.* **2015**, *52*.

39. Roberts, D.; Gee, D.G. An introduction to the structure of the Scandinavian Caledonides. *Caledonide Orogen–Scand. Relat. Areas* **1985**, *1*, 55–68.

40. Andréasson, P.-G.; Gorbatschev, R. Metamorphism in extensive nappe terrains: A study of the Central Scandinavian Caledonides. *Geologiska Föreningen i Stockholm Förhandlingar* **1980**, *102*, 335–357. [CrossRef]

41. Be'Eri-Shlevin, Y.; Gee, D.; Claesson, S.; Ladenberger, A.; Majka, J.; Kirkland, C.; Robinson, P.; Frei, D. Provenance of Neoproterozoic sediments in the Särv nappes (Middle Allochthon) of the Scandinavian Caledonides: LA-ICP-MS and SIMS U–Pb dating of detrital zircons. *Precambrian Res.* **2011**, *187*, 181–200. [CrossRef]

42. Gee, D.G.; Fossen, H.; Henriksen, N.; Higgins, A.K. From the Early Paleozoic Platforms of Baltica and Laurentia to the Caledonide Orogen of Scandinavia and Greenland. *Episodes* **2008**, *31*, 44–51. [CrossRef] [PubMed]

43. Ladenberger, A.; Be'eri-Shlevin, Y.; Claesson, S.; Gee, D.G.; Majka, J.; Romanova, I.V. Tectonometamorphic evolution of the Åreskutan Nappe–Caledonian history revealed by SIMS U–Pb zircon geochronology. *Geol. Soc. Lond. Spec. Publ.* **2014**, *390*, 337–368. [CrossRef]

44. Gromet, L.; Sjöström, H.; Bergman, S.; Claesson, S.; Essex, R.; Andréasson, P.; Albrecht, L. Contrasting ages of metamorphism in the Seve nappes: U-Pb results from the central and northern Swedish Caledonides. *GFF* **1996**, *118*, 36–37. [CrossRef]

45. Williams, I.S.; Claesson, S. Isotopic evidence for the Precambrian provenance and Caledonian metamorphism of high grade paragneisses from the Seve Nappes, Scandinavian Caledonides. *Contrib. Mineral. Petrol.* **1987**, *97*, 205–217. [CrossRef]

46. Bender, H.; Glodny, J.; Ring, U. Absolute timing of Caledonian orogenic wedge assembly, Central Sweden, constrained by Rb–Sr multi-mineral isochron data. *Lithos* **2019**, *344–345*, 339–359. [CrossRef]

47. Klonowska, I.; Janák, M.; Majka, J.; Petrík, I.; Froitzheim, N.; Gee, D.G.; Sasinková, V. Microdiamond on Åreskutan confirms regional UHP metamorphism in the Seve Nappe Complex of the Scandinavian Caledonides. *J. Metamorph. Geol.* **2017**, *35*, 541–564. [CrossRef]

48. Tsang, C.-F.; Rosberg, J.-E.; Sharma, P.; Berthet, T.; Juhlin, C.; Niemi, A. Hydrologic testing during drilling: Application of the flowing fluid electrical conductivity (FFEC) logging method to drilling of a deep borehole. *Hydrogeol. J.* **2016**, *24*, 1333–1341. [CrossRef]

49. Kamber, B.S.; Whitehouse, M.J. Micro-scale sulphur isotope evidence for sulphur cycling in the late Archean shallow ocean. *Geobiology* **2007**, *5*, 5–17. [CrossRef]

50. Ding, T.; Valkiers, S.; Kipphardt, H.; De Bièvre, P.; Taylor, P.; Gonfiantini, R.; Krouse, R. Calibrated sulfur isotope abundance ratios of three IAEA sulfur isotope reference materials and V-CDT with a reassessment of the atomic weight of sulfur. *Geochim. Cosmochim. Acta* **2001**, *65*, 2433–2437. [CrossRef]

51. Roberts, N.M.W.; Rasbury, E.T.; Parrish, R.R.; Smith, C.J.; Horstwood, M.S.A.; Condon, D.J. A calcite reference material for LA-ICP-MS U-Pb geochronology. *Geochem. Geophys. Geosystems* **2017**, *18*, 2807–2814. [CrossRef]

52. Ludwig, K. Isoplot version 4.15: A geochronological toolkit for microsoft Excel. *Berkeley Geochronol. Cent. Spec. Publ.* **2008**, *4*, 247–270.

53. Hill, C.A.; Polyak, V.J.; Asmerom, Y.; Provencio, P.P. Constraints on a Late Cretaceous uplift, denudation, and incision of the Grand Canyon region, southwestern Colorado Plateau, USA, from U-Pb dating of lacustrine limestone. *Tectonics* **2016**, *35*, 896–906. [CrossRef]

54. Horstwood, M.S.A.; Kosler, J.; Gehrels, G.; Jackson, S.E.; McLean, N.M.; Paton, C.; Pearson, N.J.; Sircombe, K.; Sylvester, P.; Vermeesch, P.; et al. Community-Derived Standards for LA-ICP-MS U-(Th-)Pb Geochronology—Uncertainty Propagation, Age Interpretation and Data Reporting. *Geostand. Geoanalytical Res.* **2016**, *40*, 311–332. [CrossRef]

55. Peckmann, J.; Thiel, V. Carbon cycling at ancient methane–seeps. *Chem. Geol.* **2004**, *205*, 443–467. [CrossRef]

56. Sandström, B.; Tullborg, E.-L. Episodic fluid migration in the Fennoscandian Shield recorded by stable isotopes, rare earth elements and fluid inclusions in fracture minerals at Forsmark, Sweden. *Chem. Geol.* **2009**, *266*, 126–142. [CrossRef]

57. Campbell, K.; Farmer, J.D.; Marais, D.D. Ancient hydrocarbon seeps from the Mesozoic convergent margin of California: Carbonate geochemistry, fluids and palaeoenvironments. *Geofluids* **2002**, *2*, 63–94. [CrossRef]

58. Tullborg, E.-L.; Landström, O.; Wallin, B. Low-temperature trace element mobility influenced by microbial activity—Indications from fracture calcite and pyrite in crystalline basement. *Chem. Geol.* **1999**, *157*, 199–218. [CrossRef]

59. Clauer, N.; Frape, S.K.; Fritz, B. Calcite veins of the Stripa granite (Sweden) as records of the origin of the groundwaters and their interactions with the granitic body. *Geochim. Cosmochim. Acta* **1989**, *53*, 1777–1781. [CrossRef]

60. Canfield, D. Isotope fractionation by natural populations of sulfate-reducing bacteria. *Geochim. Cosmochim. Acta* **2001**, *65*, 1117–1124. [CrossRef]

61. Ries, J.B.; Fike, D.A.; Pratt, L.M.; Lyons, T.W.; Grotzinger, J.P. Superheavy pyrite (δ34Spyr> δ34SCAS) in the terminal Proterozoic Nama Group, southern Namibia: A consequence of low seawater sulfate at the dawn of animal life. *Geology* **2009**, *37*, 743–746. [CrossRef]

62. Jones, D.S.; Fike, D.A. Dynamic sulfur and carbon cycling through the end-Ordovician extinction revealed by paired sulfate–pyrite δ34S. *Earth Planet. Sci. Lett.* **2013**, *363*, 144–155. [CrossRef]

63. Sim, M.S.; Bosak, T.; Ono, S. Large Sulfur Isotope Fractionation Does Not Require Disproportionation. *Sci.* **2011**, *333*, 74–77. [CrossRef] [PubMed]

64. Wortmann, U.G.; Bernasconi, S.M.; Böttcher, M.E. Hypersulfidic deep biosphere indicates extreme sulfur isotope fractionation during single-step microbial sulfate reduction. *Geology* **2001**, *29*, 647. [CrossRef]

65. Drake, H.; Tullborg, E.-L.; Whitehouse, M.; Sandberg, B.; Blomfeldt, T.; Åström, M.E. Extreme fractionation and micro-scale variation of sulphur isotopes during bacterial sulphate reduction in deep groundwater systems. *Geochim. Cosmochim. Acta* **2015**, *161*, 1–18. [CrossRef]

66. Canfield, D.E.; Farquhar, J.; Zerkle, A.L. High isotope fractionations during sulfate reduction in a low-sulfate euxinic ocean analog. *Geology* **2010**, *38*, 415–418. [CrossRef]

67. Laaksoharju, M.; Smellie, J.; Tullborg, E.-L.; Gimeno, M.; Molinero, J.; Gurban, I.; Hallbeck, L. Hydrogeochemical evaluation and modelling performed within the Swedish site investigation programme. *Appl. Geochem.* **2008**, *23*, 1761–1795. [CrossRef]

68. Machel, H. Bacterial and thermochemical sulfate reduction in diagenetic settings—Old and new insights. *Sediment. Geol.* **2001**, *140*, 143–175. [CrossRef]

69. Goldstein, R.H. Fluid inclusions in sedimentary and diagenetic systems. *Lithos* **2001**, *55*, 159–193. [CrossRef]

70. Johansson, E.; Gustafsson, L.-G.; Berglund, S.; Lindborg, T.; Selroos, J.-O.; Liljedahl, L.C.; Destouni, G. Data evaluation and numerical modeling of hydrological interactions between active layer, lake and talik in a permafrost catchment, Western Greenland. *J. Hydrol.* **2015**, *527*, 688–703. [CrossRef]

71. Stuevold, L.M.; Eldholm, O. Cenozoic uplift of Fennoscandia inferred from a study of the mid-Norwegian margin. *Glob. Planet. Chang.* **1996**, *12*, 359–386. [CrossRef]

72. Pitkänen, P.; Partamies, S.; Luukkonen, A. *Hydrogeochemical Interpretation of Baseline Groundwater Conditions at the Olkiluoto Site*; Posiva Olkiluto: Helsinki, Finland, 2004.

73. Milodowski, A.E.; Bath, A.; Norris, S. Palaeohydrogeology using geochemical, isotopic and mineralogical analyses: Salinity and redox evolution in a deep groundwater system through Quaternary glacial cycles. *Appl. Geochem.* **2018**, *97*, 40–60. [CrossRef]

74. Bird, D.K.; Spieler, A.R. Epidote in geothermal systems. *Rev. Mineral. Geochem.* **2014**, *56*, 235–300. [CrossRef]

75. Liou, J.G.; Maruyama, S.; Cho, M. Phase equilibria and mixed parageneses of metabasites in lowgrade metamorphism. *Mineral. Mag.* **1985**, *49*, 321–333. [CrossRef]

76. Dennis, P.F.; Myhill, D.J.; Marca, A.; Kirk, R. Clumped isotope evidence for episodic, rapid flow of fluids in a mineralized fault system in the Peak District, UK. *J. Geol. Soc.* **2019**, *176*, 447–461. [CrossRef]

Article

A Novel Approach to Isolation and Screening of Calcifying Bacteria for Biotechnological Applications

Paola Cacchio * and Maddalena Del Gallo

Department of Life, Health & Environmental Sciences, Microbiology Laboratory, University of L'Aquila, Coppito, 67010 L'Aquila, Italy; mariamaddalena.delgallo@univaq.it
* Correspondence: paola.cacchio@univaq.it

Received: 31 October 2019; Accepted: 12 November 2019; Published: 14 November 2019

Abstract: Bacterial calcium-carbonate precipitation (BCP) has been studied for multiple applications such as remediation, consolidation, and cementation. Isolation and screening of strong calcifying bacteria is the main task of BCP-technique. In this paper, we studied $CaCO_3$ precipitation by different bacteria isolated from a rhizospheric soil in both solid and liquid media. It has been found, through culture-depending studies, that bacteria belonging to *Actinobacteria*, *Gammaproteobacteria*, and *Alphaproteobacteria* are the dominant bacteria involved in $CaCO_3$ precipitation in this environment. Pure and mixed cultures of selected strains were applied for sand biocementation experiments. Scanning electron microscopy (SEM) and energy dispersive X-ray spectroscopy (EDS) analyses of the biotreated samples revealed the biological nature of the cementation and the effectiveness of the biodeposition treatment by mixed cultures. X-ray diffraction (XRD) analysis confirmed that all the calcifying strains selected for sand biocementation precipitated $CaCO_3$, mostly in the form of calcite. In this study, Biolog® EcoPlate is evaluated as a useful method for a more targeted choice of the sampling site with the purpose of obtaining interesting candidates for BCP applications. Furthermore, ImageJ software was investigated, for the first time to our knowledge, as a potential method to screen high $CaCO_3$ producer strains.

Keywords: bacterial calcium-carbonate precipitation (BCP); calcifying bacteria selection; calcifying mixed cultures; ImageJ software; Biolog EcoPlates; sand biocementation

1. Introduction

In natural environments, chemical $CaCO_3$ precipitation is accompanied by microbial processes involving bacteria, particularly cyanobacteria, archaea, small algae, and fungi [1]. Bacterial calcium-carbonate precipitation (BCP) is of major ecological and geological importance, and it has been widely studied in soils, caves, lakes, spring, and seawater, as well as in the laboratory [2–15]. Chemical $CaCO_3$ precipitation is a rather straightforward process governed mainly by concentration of non-precipitated calcium, concentration of the total inorganic carbon, pH, and availability of nucleation sites for $CaCO_3$ crystal formation [16–18]. Bacteria can alter these factors, either separately or in combinations with one another, in a number of ways linked to cell surface structures and metabolic activities [16–36]. For example, in super-saturated solution, bacteria can become the nucleus of mineral precipitation due to Ca^{2+} adsorption to the capsule, cellular surface membrane, cell wall, or EPS (exo-poly-saccharides) layer [21,23–26]. Alternatively, in under-saturated solution, bacteria can induce precipitation of $CaCO_3$ through metabolic pathways associated with photosynthesis [17,19,28,29,32–34], nitrogen and sulfur cycles [17,31], and ion exchange (Ca^{2+}/H^+) [20]. These different metabolic processes increase pH and/or dissolved inorganic carbon concentration, favoring $CaCO_3$ precipitation.

The various ecological implications of BCP in biotechnology have been described by Zhu and Dittrich [37]. BCP applications include: Bioremediation of metal contaminated soil and ground water; restoration and preservation of calcareous sculptural artefacts, historical monuments and civil buildings; bioconcrete; different geotechnical-engineering applications such as soil/sand strength, sand impermeability, mitigation of liquefaction, soil erosion control, impermeabilization of polluted soils and other soil improvement projects; microbial enhanced oil recovery (EOR); CO_2 sequestration; filler for rubber, plastic, and ink [38–41].

There are two ways to apply BCP technology to sand/soil improvement. The first (biostimulation) involves increasing BCP naturally occurring in the soil, through specific nutrient injection until it reaches the desired level. To date, because of its simplicity, the most commonly studied system of applied BCP is urea hydrolysis via the enzyme urease. The second (bioaugmentation) involves the use of selected calcifying bacteria that, *in primis*, produce $CaCO_3$ crystals at high amount and speed, in suitable environment. Until now, calcifying bacteria for the different biotechnological uses have been isolated mainly from soil, freshwater and seawater, different $CaCO_3$ deposits such as stalactites and stalagmites in caves, and screened based mostly on traditional method (which requires a large amount of reagent and time) [41]. However, there are few studies on the rhizosphere bacteria ability to precipitate $CaCO_3$ [27,42] despite the fact that root exudates include amino acids (whose metabolization produces the pH conditions necessary for BCP), as well as low-molecular weight organic acids whose mineralization by heterotrophic bacteria can lead to $CaCO_3$ precipitation in a circumneutral environment rich in dissolved calcium [20,27].

In this study, we enriched a consortium of bacteria isolated from rhizosphere soil based on their ability to precipitate $CaCO_3$. Then, we isolated several individual members of this community, and analyzed their metabolism and biomineralization to screen strong $CaCO_3$ depositor strains. The best strains were further characterized based on their ability to cement sand.

The novelty of this study lies above all in the microbiological approach to biocementation. It is based on the study of the indigenous microbial community of a rhizospheric soil (such as in biostimulation) and then on the use of calcifying bacterial strains or their consortia (such as in bioaugmentation) isolated from it. This is because we attempted to optimize the choice of the sampling site by Biolog EcoPlate assay (assessing the carbonatogenic potential of the microbial community dwelling the soil) to obtain effective calcifying strains to be used also in other sites, as well as for biotechnological applications other than biocementation.

Biolog EcoPlates are 96-well plates, containing in triplicate 31 different substrates plus a control [43]. The substrates in the wells can be subdivided into five groups: Carbohydrates (n = 10), carboxylic acids (n = 9), amines and amides (n = 2), amino acids (n = 6), and polymers (n = 4). These compounds occur naturally in the soil and most of them are exudate by plant root in the rhizosphere [44]. In each well there is a redox indicator (a tetrazolium salt) that changes from colorless into purple in the presence of metabolic activity of the microbial community. The intensity of the coloring, which is optically measured at 590 nm, is proportional to the substrate utilization.

As to methodological novelty, we proposed the use of ImageJ [45,46] as a low cost and rapid method to assess the in vitro mineralization aptitude of calcifying strains. A calcifying strain is of interest from a biotechnological point of view, mainly, when it produces a large amount of crystals that can be evaluated, for example, by recovering the crystals from agar and weighing them [22]. This operation is time consuming and not always leads to the expected result because crystals can remain embedded in the microbial biofilm. The ImageJ, already used in many scientific applications, allows screening in a shorter time.

2. Materials and Methods

2.1. Soil Sampling, Biolog EcoPlate Assay, and Physico–Chemical Analyses

In this study, we used the Biolog EcoPlates to assess the carbonatogenic potential of the microbial communities dwelling the rhizosphere soil of *Pinus strobus* (Pinaceae), with the purpose of choosing a proper sampling site for an efficient screening of calcifying bacterial strains (since bacterial mineralization of organic acids and amino acids are important metabolic pathways involved in BCP). As the probability to select strong calcifying bacteria should likely increase starting the screening program from a soil dwelled by metabolically active, versatile, and highly diversified microbial community, we used the Biolog EcoPlate assay to obtain information also in this regard.

The rhizosphere soil was sampled from the garden of a residential area located near L'Aquila (Central-Italy) at a depth of 5 cm, using sterile tools. The composite sample, placed into a sterile container, was carried to the laboratory at room temperature and immediately subjected to Biolog EcoPlate assay (BIOLOG Inc., Hayward, CA, USA) and microbiological analyses, performed in triplicate. 3 g of soil were added to 27 mL of sterile physiological solution (0.9% NaCl) into a sterile flask, stirred for 30 min at room temperature and then centrifuged for 2 min at 2000 rpm. The supernatant was collected and placed into a sterile Petri dish, under sterile condition. A multichannel pipette was used to inoculate 150 µL of the bacterial suspension into each well of the Biolog EcoPlate. The microplates were placed in a sterile plastic envelope, hermetically sealed, in order to limit the dehydration of the wells and incubated at 28 °C. The metabolic fingerprints were analyzed by measuring the absorbance at the Biolog Microstation, immediately following the inoculation procedure, considered as time zero, then every 24 h until 216 h.

Microbial activity in each EcoPlate, was expressed as average well color development (AWCD), which gives the index of the total metabolic potential of the community [47]. AWCD was determined as follows Equation (1):

$$\text{AWCD} = \sum_{i=1}^{31} \frac{ODi}{31} \tag{1}$$

To calculate the AWCD at each reading time, the optical densities from each well (OD_i) were corrected by subtracting the initial reading at time zero of the same well, as well as of the blank one (inoculated but without a carbon source). $OD_{590} = 0.25$ was assumed as a threshold value, below which a substrate was considered as unmetabolized [47].

The following metabolic-ecological indexes were calculated based on the ODs at 120 h, when the community reached the plateau: (i) Metabolic richness index (R) as the number of oxidized substrates, that indicates the metabolic versatility of the community [48]; (ii) the Shannon–Weaver index (H'), that measures the metabolic diversity of the species present. It was calculated as follows Equation (2):

$$H' = -\sum_{i=1}^{R} pi\,(lnpi) \tag{2}$$

where p_i (the proportional color development of the well over total color development of all wells in the plate) is calculated as the ratio of the corrected absorbance value of each well (ODi) to the sum of the absorbance value ($\sum OD_i$) of all wells in the plate [47] and (iii) the evenness index (E). This index, calculated as H'/lnR, measures the variation in color development among wells and indicates the uniformity in the microbial growth on the different substrates, which can be related to a similar abundance of the different species [47–49].

For a better knowledge of the material, physico–chemical properties of the sampled soil including pH, organic carbon content (%), total nitrogen (%), and assimilable phosphorous concentrations (mg/Kg), cation exchange capacity—CEC (meq) and exchangeable Ca^{2+}, Mg^{2+}, K^+, Na^+ (mg/Kg), total, and active calcium carbonate content (%), electric conductivity—EC (mS), texture, sand, silt, and clay fractions (%) were determined according with the official methods [9].

As shown in Table 1, the sampled soil was a sandy-loam soil containing an average amount of total carbonates and a very high amount of Ca^{2+}. Total pH was weak alkaline. The content of total N and available P were low and very low, respectively.

Table 1. Physico–chemical properties of the soil sampled for the isolation of calcifying bacteria.

Skel.	Sand (%)	Silt (%)	Clays (%)	Text.	O.M. (%)	pH	E.C. (mS)	Tot. CaCO$_3$ (%)	Act. CaCO$_3$ (%)	N (%)	Ca (ppm)	Mg (ppm)	K (ppm)	Na (ppm)	P (ppm)	Fe (ppm)	Mn (ppm)	Cu (ppm)	Zn (ppm)	C.E.C. (meq)	Ca (meq)	Mg (meq)	K (meq)	Na (meq)	B. S. (%)	Mg/K
Con.	48	15	37	C.S.	1.06	7.6	0.403	16.4	3.0	0.07	4100	244	203	66	6	9.6	13.0	1.9	1.2	23.35	20.50	2.04	0.52	0.29	100	3.9

Skel. = Skeleton; Con. = considerable (150–350 g/kg); Text. = texture; C.S. = clay, sand; O.M. = organic matter; E.C. = electric conductivity; Tot. CaCO$_3$ = total calcium carbonate content; Act. CaCO$_3$ = active calcium carbonate content; C.E.C. = cation exchange capacity; B.S. = basic saturation. Sand, silt, clay, total, and active calcium carbonate values are wt%.

2.2. Isolation of CaCO$_3$ Producing Bacteria

An amount of sample corresponding to 10 g of dried soil was resuspended in 90 mL sterile physiological solution and stirred for 50 min to detach the bacteria, whose external surface is negatively charged, from soil cations. Triplicate B4 agar plates [4] (2.5 g calcium acetate, 4.0 g yeast extract, 10.0 g glucose, and 18.0 g Biolife agar per liter of distilled water, the final pH was adjusted to 8.0 with NaOH) were inoculated with soil dilutions ranging from 10^{-2}–10^{-6} and incubated aerobically at 28 °C for 24 or 48 h (depending on the growth rate of the bacteria). B4 is an enrichment medium for calcifying bacteria; the acetate ion is an energy source, while the Ca^{2+} cation is used by the calcifying bacteria to precipitate calcium carbonate. The presence of glucose speeds the process. After its first description in 1973 [4], B4 medium was utilized by several authors [50–54] to stimulate the growth of calcifying bacterial species. The rationale is that calcifying heterotrophic bacteria belong to various groups with different nutritional and physiological properties so that the presence of nutrients widespread in the environment, such as glucose and factors present in yeast extract, is expected to enlarge the platform of enriched strains. Colonies were assessed every 24/48 h with optical microscopy (Leitz-Biomed) and selected as positive based on visual crystal formation within 25 days. Positive isolates were purified by repeated streaking on B4 agar plates and preliminarily characterized.

A serial dilution of the soil sample was also plated in triplicate on nutrient agar (Difco) for colony forming unit (CFU) counting.

Calcifying bacteria were preserved by various methods. For short-term preservation, strains were incubated at 28 °C for 2–5 days and then kept at 4 °C on slants of B4. For long-term maintenance pure cultures were kept in glycerol at −80 °C.

2.3. Preliminary Characterization and Biolog Identification

Morphological characterization of the bacterial colony was performed on the basis of pigmentation, form, elevation, margin, opacity, and surface [55]. Optical microscope (Leitz-Biomed, 40× and 100×) was used to establish the shape of the heat-fixed bacterial cells after Gram staining by the Color Gram 2 Kit (bioMérieux, Marcy-l'Etoile/France), following the kit instructions. Physiological characterization was performed on the basis of the relation to temperature (bacterial isolates were incubated on B4 agar at 4, 20, and 28 °C, significant temperatures for in situ biocementation experiments) and oxygen presence (since anaerobic conditions are abundant in the soil, particularly when the rain saturates the soil and gas exchange slows down almost completely) by incubating bacterial strains on B4 agar in an anaerobic chamber (Oxoid, Hampshire, UK). The occurrence of a catalase reaction was demonstrated by the formation of bubbles after mixing a suspension of one-day bacterial cultures with a drop of 3% (v/v) hydrogen peroxide water-solution on slide. Urease activity of calcifying strains was detected on Christensen agar medium containing 2% urea and phenol red as pH indicator [56,57]. We cooled the sterilized medium (115 °C for 20 min) to 40–50 °C, before adding the urea supplement. Inoculated agar slants were incubated at 28 °C for 24/48 h and observed for color change after 6, 24, and 48 h of incubation. Microorganisms with urease activity hydrolyze urea with the ammonium ions production causing a change of color of the agar medium to fuchsia (alkaline pH).

Biolog GN2 and GP2 plates (Biolog, Inc., Hayward, Calif., USA) were used for the identification of Gram-negative and Gram-positive bacteria, respectively. One or two days (depending on the growth rate of the bacterium) before the inoculation on Biolog GN2 and GP2 plates, bacterial strains were streaked on Biolog universal growth agar (BUG) and incubated at 28 °C. Inoculation and reading of micro-plates were carried out according to the instructions of manufacturer using a Biolog Microstation with OmniLog. The Biolog method is based on oxidation tests of 93 substrates in a 96-well microtiter plate. Each well contains a redox dye, tetrazolium violet, that allows colorimetric determination of the increased respiration that occurs when microbial cells are oxidizing a C-substrate. Reactions were compared to the GP or GN database based on the similarity index which must be at least 0.75 to be considered an acceptable species identification after 4 h of incubation and at least 0.50 after 16–24 h of incubation.

2.4. Calcification and CaCO₃ Solubilization Activities

Pure bacterial cultures on B4 agar plates were observed every 24/48 h under a light microscope (Leitz-Biomed, 10×) for up to 25 days after inoculation, to follow crystal production by the bacterial strains at different temperatures, under aerobic condition. Three temperatures 4, 20, and 28 °C were selected to test the induction of carbonate precipitation by the bacterial isolates, as representative of natural conditions. The bacterial aptitude for calcification was evaluated on the basis of: (i) $CaCO_3$ crystal amount on B4 agar, (ii) size of the $CaCO_3$ bioliths, (iii) presence of $CaCO_3$ aggregates, (iv) time to initiate the precipitation activity. Position of crystals on B4 agar (on the colony, in proximity, or far away in the medium) was also checked.

Calcifying bacterial strains were inoculated also into 100 mL of sterile B4 liquid medium, to assess their capability to precipitate in liquid conditions. Inoculated flasks were incubated under static conditions at 28 °C for two months. The presence of crystals was macroscopically assessed both at the bottom of the flasks and on the wall.

As microbial mediated reactions can generate considerable amounts of H^+ ions that can dissolve $CaCO_3$ deposits, we tested the ability of the selected calcifying strains to dissolve $CaCO_3$ after 7, 15, and 30 days of incubation at 28 °C. Calcifying isolates were grown on the Deveze-Bruni medium (1.0 g peptone, 1.0 g yeast extract, 5.0 g glucose, 5.0 g NaCl, 0.4 g K_2HPO_4, 0.5 g $(NH_4)_2SO_4$, 0.1 g $MgCl_2$, 0.01g $FeCl_2$, 0.1 g $CaCl_2$, pH = 6.8) supplemented with 0.14 or 2.5% (w/v) $CaCO_3$ [58]. Solubilization activity was shown as clear halo that surrounded each colony in response to decreased pH [59].

2.5. ImageJ Analysis

In the present research, the ImageJ software was evaluated as a rapid screening method to estimate the extent of $CaCO_3$ production on B4 agar plates by each calcifying bacterial strain, after 30 days of growth at 28 °C. ImageJ analysis was performed from digital images of five different areas (total area = 125 × 5 mm²) of the inoculated B4 agar plates (plate area = 6359 mm²) observed upside down at optical microscope (Leitz-Biomed, 10×) (Figure 1). These areas were randomly chosen and acquired through an iPhone, used as the camera, directly from the eyepiece of the microscope. By ImageJ analysis we calculated the total area occupied by the crystals on each optical image and the mean (with its standard deviation) on each set of the five images, to obtain the percent average surface covered by $CaCO_3$ crystals (referred as BCP extent) deposited in vitro by each calcifying strain. In this preliminary study, the scale bar was determined by a micrometric slide.

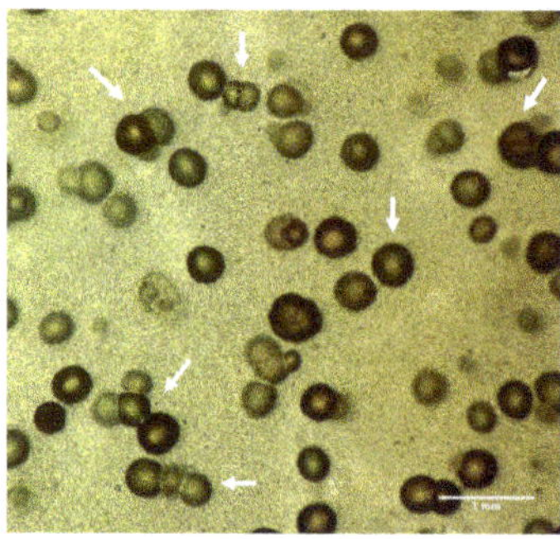

Figure 1. A digital image acquired through an iPhone (used as the camera) of an inoculated B4 agar plate, observed upside down at optical microscope (Leitz-Biomed, 10×). Note globular $CaCO_3$ crystals of different dimensions and their aggregates precipitated by the strain P1 (**arrows**), after 30 days of growth at 28 °C.

2.6. Sand Biocementation Test

Experiments of consolidation by BCP were carried out using sand samples from the Adriatic-sea coast (Central Italy) and bacterial calcifying cultures selected during this study. Table 2 shows some of the physical and chemical characteristics of the sand used in this study.

Table 2. Physico–chemical properties of the sea sand utilized for biocementation experiments.

Skel.	Sand (%)	Silt (%)	Clay (%)	Text.	O.M. (%)	pH	E.C. (mS)	Tot. CaCO$_3$ (%)	Act. CaCO$_3$ (%)	N (%)	Ca (ppm)	Mg (ppm)	K (ppm)	Na (ppm)	P (ppm)	Fe (ppm)	Mn (ppm)	Cu (ppm)	Zn (ppm)	C.E.C. (meq)	Ca (meq)	Mg (meq)	K (meq)	Na (meq)	B. S. (%)	Mg/K
TRA.	94	2	4	S.	0.04	7.9	0.57	33.9	0.9	0.009	2940	77	54	119	6	3.6	4.6	0.4	0.8	16.01	14.7	0.65	0.14	0.52	100	4.6

Skel. = Skeleton; TRA. = trace; Text. = texture; S. = sandy; O.M. = organic matter; E.C. = electric conductivity; Tot. CaCO$_3$ = total calcium carbonate content; Act. CaCO$_3$ = active calcium carbonate content; C.E.C. = cation exchange capacity; B.S. = basic saturation. Sand, silt, clay, total, and active calcium carbonate values are wt %.

The sand was characterized by a weak alkaline pH, a medium content of total carbonates and a high level of exchangeable Ca^{2+}. The content of organic matter as well as of total N and assimilable P were very low.

The following biocementation experiments were set up: (i) 7 g of sand, on Petri dish, for three weeks; (ii) 300 g of sand, in Magenta vessel, for two months; (iii) 300 g of sand, in Magenta vessel, for two months in the presence of equimolar (1 M) urea and calcium chloride, as calcification inducing-agents. Control specimens were sand samples treated with not inoculated B4 medium.

For sand biocementation on Petri dish, strain selection was based on: (i) Calcification extent (% area covered by $CaCO_3$ crystals precipitated on B4 agar plates) established by ImageJ software. This parameter depends on amount and size of the precipitated $CaCO_3$ crystals, presence of $CaCO_3$ aggregates, days to initiate bio-precipitation, $CaCO_3$ solubilization activity; (ii) calcification parameters not included in the calcification extent, such as biodeposition in liquid medium and capability to precipitate $CaCO_3$ at different temperatures; (iii) cell morphology; (iv) relation with oxygen; (v) Gram-staining 7 g of washed sand, sterilized at 120 °C for 20 min and dried in an oven, were homogenously distributed on a 90 mm Petri dish and then inoculated with 25 mL of a calcifying cell suspension (10^7 CFU/mL), grown on B4 medium. Biopretreated samples were incubated at 28 °C statically, to allow the bacteria to cement the sand.

To test BCP in depth, 300 g of sand were placed in a Magenta vessel and inoculated with 150 mL of a calcifying cell suspension (10^6 CFU/mL), grown on B4 medium, in the presence or not of the inducing agents (equimolar urea and calcium chloride). For sand biocementation in Magenta vessel, strain selection was based on: (i) Calcification extent (established by ImageJ software), (ii) ability to precipitate $CaCO_3$ crystals in B4 liquid medium, (iii) ability to hydrolyze urea, and (iv) growth in anaerobiosis.

2.7. SEM, XRD, and EDS Analyses

Scanning electron microscopy (SEM) was used to study crystal morphology and their relationship with microbial cells. Samples were prepared as follows: Samples on agarized media were dried at 37 °C for 40 days; agar medium was cut into flat blocks, gold-sputtered, and observed with a Philips SEM XL30CP. For crystallite-poor samples, the agar was dissolved and crystals were collected and purified according to the method of Rivadeneyra [22].

XRD was used to identify $CaCO_3$ mineral phases of the bioprecipitated crystals. XRD analyses were done by a two-circle θ/2θ diffractometer with a Cu radiation source (Bruker D5000). The supply voltage of the X-ray tube was set at 40 kV, 40 mA. The 2θ scan range was between 20 and 60°; each scan was done in steps of 0.02. A counting time between 30 and 40 s per step was selected, depending on the sample density. The crystalline phases were identified using the ICDD database (JCPDS). XRD samples were prepared as follows: Cultured solid media were dried at 32 °C, agar medium was cut into 10 × 30 mm flat blocks, 0.5 mm high, and those richest in crystallites were fixed by adhesive tape on a glass slide for X-ray measurements.

SEM and EDS were used to detect, after 20 days-treatment, the nature of sand cementation by the calcifying cultures. For EDS, samples were dried, uniformly spread directly on adherent tabs, and analyzed without gold film. SEM-EDS analysis was performed by a SEM-Philips XL30CP and a EDX-INCA ENERGY 250.

XRD and EDS were also used to study the untreated sea sand.

3. Results and Discussion

3.1. Biolog EcoPlates Assay

Since BCP is influenced by the metabolic activity of the bacteria, we investigated through the Biolog EcoPlates the metabolic activity of the bacterial communities dwelling a rhizospheric soil, with the purpose of choosing a good sampling site for an efficient calcifying bacteria screening. In the literature there is a lot of information about the possibilities of using the Biolog EcoPlates to analyze functional microbial diversity in the presence of different plant and cultivation type, and to study the metabolic activity of microorganisms in soils contaminated with pesticides and heavy metals.

To our knowledge, Andrei et al. [60] were the first to use Biolog EcoPlates to investigate the carbonatogenic potential of bacterial communities (dwelling a limestone statue in Romania), based on their ability to metabolize amino acids since this metabolization promotes the pH conditions necessary for the biomineralization process (by ammonia released in the process of oxidative deamination). Overall, Andrei et al. study highlights the need to evaluate the carbonatogenic potential of the bacterial communities present on a stone artwork prior to designing an efficient conservation treatment based on resident calcifying bacteria. On the same line, we used the Biolog EcoPlate assay to assess the carbonatogenic potential of the bacterial community dwelling a rhizospheric soil at the aim of starting an efficient screening program of calcifying bacteria.

Unlike Andrei et al., in this study we have correlated the carbonatogenic potential of the bacterial communities to the metabolic use not only of the amino acids but also of the organic acids present in the Biolog EcoPlate. This is because both these carbon compounds occur naturally in the rhizospheric soil environment, as plant root exudates, as well as because low-molecular weight organic acids mineralization by heterotrophic bacteria is another microbial process that (increasing both pH and concentration of dissolved inorganic carbon) can lead to calcium carbonate precipitation in a circumneutral environment rich in dissolved calcium. As $CaCO_3$ precipitation always appears to be a response of the heterotrophic bacterial communities to an enrichment of the environment in organic matter in aerobiosis, anaerobiosis, and microaerophilia, this mineralization process has been commonly used in microbial carbonate precipitation experiments [17,20]. Moreover, since the probability of obtaining efficient $CaCO_3$ producer strains increases likely in the presence of bacterial communities highly biodiversified with a high catabolic versatility, we used the Biolog EcoPlate assay also to obtain information about these metabolic and ecological aspects. Based on these considerations, all the information of the studied rhizospheric soil obtained through the Biolog EcoPlate system revealed its usefulness to start a successful screening of calcifying bacteria. In fact, we found that at the end of the incubation period, the soil microbial community used 29 out of 31 different substrates (metabolic Richness index $R = 29$) present in the Biolog EcoPlate, including all the amino acids and almost all the organic acids (apart from 2-hydroxybenzoic acid and α-ketobutyric acid). This means that the microbial community under study had a high functional diversity and catabolic versatility as well as a good mineralization potential of $CaCO_3$ (the utilization of all the six amino acids, as well as of almost all the carboxylic acids indicates the presence in the soil of favorable conditions for the growth of microorganisms carrying the enzymes taking part in metabolic pathways of BCP). The general catabolic activity of the soil community to utilize the C-substrates, turned out to be not only wide (as shown by the metabolic Richness index, R) but also high as shown by the dynamic of the AWCD during the incubation time of the EcoPlates, at 28 °C (Figure 2).

Figure 3 showed that microbial functional diversity changed over time, and that the microbial community initially prefer C-substrates from the carboxylic and amino acid groups (namely D-Galacturonic acid and L-Asparagine), in addition to Glucose-1-phosphate. During the exponential phase, an increasing number of substrates (belonging to all five groups) was utilized (Figure 3). However, the diversification over time of the microbial substrate use from the different groups (polymers, carbohydrates, carboxylic acids, amino acids, and amines) is beyond the aim of this research and will not be further discussed.

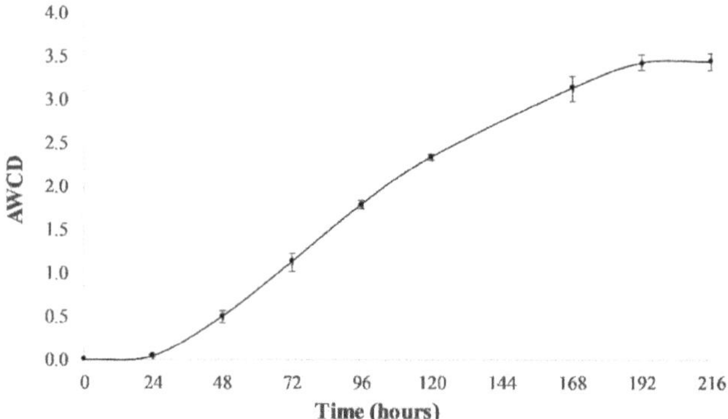

Figure 2. Dynamic of the average well color development (AWCD) of the rhizospheric-soil sample during the incubation time (9 days), at 28 °C (AWCD points are means of three replicates). Data were corrected subtracting the initial readings (at 590 nm) at time zero, as well as the reading of the control well (inoculated but without a C-substrate). The curve shows how the microbial community had a long lag-time in the development of formazan products within the Biolog-plate wells. Development was consistent and linear over the entire incubation period (9 days). The number of positive wells increased slowly for the first 24 h and then increased up to 192 h of incubation, when stationary phase began.

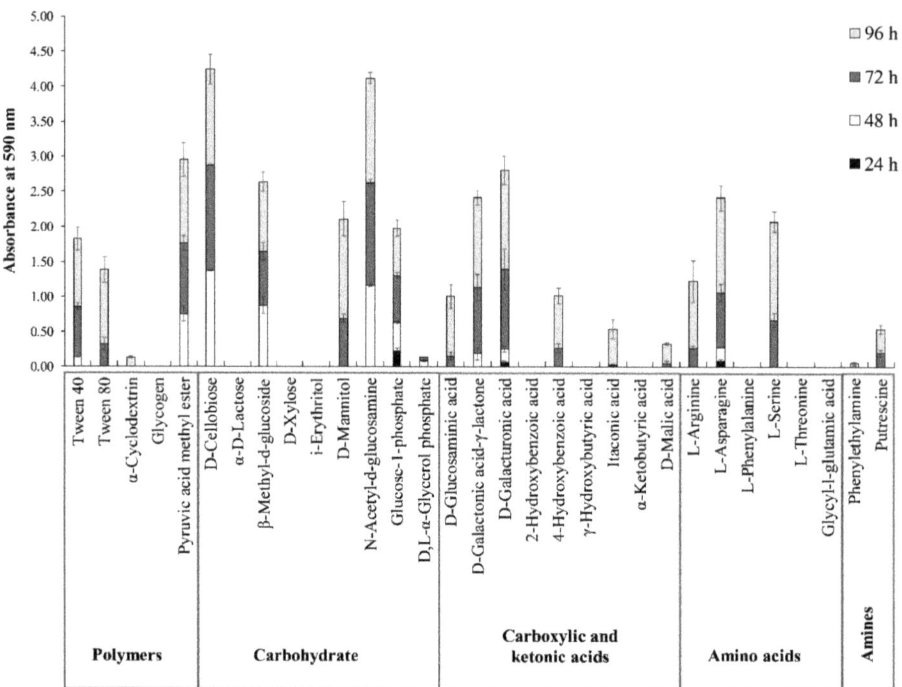

Figure 3. Stacked columns showed the diversification over time of the substrate use from the different groups (polymers, carbohydrates, carboxylic and ketonic acids, amino acids and amines) by the soil microbial community. In the figure, absorbances are reported (recorded at 590 nm) for each substrate from 0 to 48 h (from 72 to 96 h, and from 96 to 216 h. The chart shows how the substrate use in the first 24 h (during the lag-phase, Figure 2), is limited to Glucose-1-phosphate, L-Asparagine, and D-Galacturonic acid. During the exponential phase, an increasing number of substrates (belonging to all five groups) was used by the microbial community.

The substrate uniformity index (*E*) value was 0.95, highlighting a high uniformity in the activity of the different substrates (that can be related to a similar abundance of the different species), indicating from our specific point of view, a greater chance of isolating bacteria capable to perform ammonification (of amino acids) and low-molecular weight organic acid mineralization. Finally, the Shannon–Weaver index (*H'*) was equal to 3.20, highlighting a high diversity of the species present related to the high diversity of nutrients available for growth in the root zones, including amino acids and organic acids [61].

3.2. Isolation and Preliminary Characterization of Heterotrophic Calcifying Bacterial Strains

By using traditional cultivation techniques, a discrete cultivable heterotrophic bacterial community was isolated from the sampled rhizospheric soil (4.5×10^5 CFU/g dw). The cultivable heterotrophic bacterial density was lower than expected (10^6–10^{10} CFU/g dw), possibly because sampling was done in winter. In a previous work [9], we also found that the cultivable microbial biomass is significantly related to the silt fraction, negatively related to the clay fraction, and negatively related to the Ca^{2+} content (Table 1). As expected, based on the weak alkaline pH value (Table 1), bacteria represented 78% of the cultivable microbial population. Fungi were the remaining component. Fourteen morphologically different heterotrophic bacteria (named from P1 to P14, where P stands for pure) were isolated from this microbial community and purified by streaking on B4 agar. Table 3 shows characterization results of the bacterial calcifying isolates.

Table 3. Morphological, physiological, and biochemical traits of the heterotrophic calcifying bacteria isolated from a rhizospheric soil near L'Aquila (Central Italy).

Strain	Cell	Gram	4 °C	20 °C	28 °C	Anaerobic Growth	Urease
P1	Short rod	+	-	+	+	-	+ +
P2	Short rod	-	-	+	+	-	+ +
P3	Short/irregular rod	+	-	+	+	-	+ + +
P4	Short/irregular rod	+	-	+	+	-	+ + +
P5	Rod	+	-	+	+	+	+
P6	Short rod	+	-	+	+	+	+ +
P7	Rod	-	-	+	+	-	+ + +
P8	Short/irregular rod	+	+	+	+	+	+ + +
P9	Short/irregular rod	+	-	+	+	+	+ +
P10	Short/irregular rod	+	-	+	+	-	+ +
P11	Short/branched rod	+	-	+	+	-	+
P12 *	Rod	-	n.d.	+	+	+	+
P13 *	Rod	-	n.d.	+	+	+	n.d.
P14 *	Rod	-	n.d.	+	+	+	-

Gram + = Gram positive; Gram − = Gram negative; Bacterial growth at 4, 20, and 28 °C in aerobiosis and in anaerobiosis at 28 °C: (-) = absent, (+) = present; urease activity (+) = trace, (+ +) = good, (+ + +) = very good; n.d.= not determined. * Strains P12, P13, and P14, from the fourth purification step of strain P5, were recovered and characterized after the loss of P5.

All the bacterial isolates formed rod-shaped cells with different length, sometimes irregular or branched. The results of morphological characterization also showed the prevalence (64%) of Gram-positive bacteria; only strains P2, P7, P12, P13, and P14 were Gram-negative. This prevalence is in line with the results of previous works [6,8]. It is known from the literature that Gram-positive bacteria are the predominant bacteria in the soil: Only coryneform bacteria of the genus Arthrobacter alone account for approximately 20%–25% of the soil cultivable bacterial population [62]. However,

Gram-negative bacteria tend to be more abundant in rhizospheric soil compared to the bulk soil. Gram-negative bacteria biomass increases when rapidly decomposable C compounds (such as sugars, organic, and amino acids) are available [63], whereas higher proportions of Gram-positive bacteria are usually found in resource-limited area [64,65].

Strain P8 showed growth capacity at all three tested temperatures, all other strains grew only at 20 and 28 °C. Strains P5, P6, P8, P9, P12, P13, and P14 showed growth also in anaerobiosis and are therefore considered facultative anaerobes. The use of facultative anaerobic bacteria is fundamental for soil biocementation *in situ*, as in the soil co-exist aerobic and anaerobic microzones [66].

All the strains were catalase positive. Apart from the strain P14 (urease negative) and P13 (not determined), all the calcifying bacteria hydrolyzed urea. This result is in accordance with the literature, since urea production, even in the presence of a significant amount of ammonium, is a common feature among soil bacteria. Urea hydrolysis producing ammonia, increases pH and leads to the production of carbonates. This means that the application of stimulated biocalcification, based on the urea hydrolysis by native or selected (in the laboratory) microbial strains, is potentially useful in a variety of soil bio-engineering applications [67].

During the characterization stage, strain P5 showed serious growth problems and was replaced by strains P12, P13, and P14, from its fourth purification step. These strains were used along this study as a mixed calcifying culture, named M5, containing in a similar proportion the strains P12, P13, and P14.

3.3. Identification of the Calcifying Bacterial Strains

By the Biolog identification system, all the Gram-positive short/pleomorphic rod isolates were identified as coryneform bacteria, and *Clavibacter agropyri* was the most common (Tables 3 and 4).

Table 4. Identification of the calcifying bacterial isolates.

Strain	Identified Species	% ID
P1	Not identified	
P2	*Vibrio tubiashii*	94
P3	*Clavibacter agropyri*	93
P4	*Clavibacter agropyri*	73
P6	*Corynebacterium urealyticum*	78
P7	Not identified	
P8	*Clavibacter agropyri*	87
P9	*Clavibacter agropyri*	73
P10	*Clavibacter agropyri*	87
P11	*Sanguibacter suarezii*	98
P12	*Sphingomonas sanguinis*	98
P13	Not identified	
P14	*Pseudomonas syringae pv persicae*	92

% ID = Identification index.

Isolates P3, P4, P8, P9, and P10, identified as *Clavibacter agropyri* (Corynebacteria), were Gram-positive rod-shaped cells, often disposed to an angle to each other giving irregular forms (Table 3). The genus *Clavibacter* contains plant pathogenic coryneform bacteria, causing gummosis, characterized by the presence of 2,4-diaminobutyric acid as a cell wall component [68–70]. Strain P6, a facultative anaerobic Gram-positive bacterium with good urease activity, was identified as *Corynebacterium urealyticum*. The aerobic Gram-positive strain P11 was identified as *Sanguibacter suarezii*, a coryneform bacterium first isolated from blood animal samples [71]. Coryneform bacteria are widespread in nature, being common inhabitants of water and soil, where they form alone approximately 20%–25% of the soil fertile cultivable bacterial population [62]. They may be pathogens or commensals of humans, animals, and plants. This ecological diversity is matched by their wide range of biochemical properties making them an interesting group of bacteria for biotechnology purposes. Among the Gram-negative isolates, strain P2 was identified by the Biolog system as *Vibrio tubiashii* (% PROB = 94). The genus *Vibrio*, includes a great diversity of species that are common inhabitants of the aquatic environment where they are usually closely associated with many kinds of marine organisms [72]. *V. tubiashii* is a halophilic species, firstly isolated by Tubiash et al. [73], pathogenic for the larvae of bivalve mollusks. Its genome consists of two chromosomes, two megaplasmids and two plasmids. This might be a survival strategy acquired by the *Vibrios* that has facilitated their adaptation to various niches, including soil environments, in the evolution process [74,75]. Strains P12 and P14 were identified as *Sphingomonas sanguinis* and *Pseudomonas syringae*, respectively. Due to their biodegradative and biosynthetic capabilities, both of them have been used for a wide range of biotechnological applications. Strains P1, P13, and P7 were not identified (Table 4). The composition of the cultivable component of the microbial community living in the rhizospheric soil of *Pinus strobus*, showed the presence of a wide range of bacteria from diverse phylogenetic affiliation mainly *Actinobacteria*, *Alphaproteobacteria*, and *Gammaproteobacteria*. No other attempts were made to identify the isolated bacterial species as our attention was directed towards the optimization of the process stage for isolating and screening of effective calcifying bacteria.

3.4. Calcification and CaCO$_3$ Solubilization Activities

Observation of pure solid bacterial cultures under light microscopy, showed that 100% of the bacterial isolates were capable of forming crystalline CaCO$_3$ on B4 agar plates at 20 and 28 °C (Table 5). This result confirms that: (i) Biomineralization leading to CaCO$_3$ deposits is a quite common phenomenon in natural habitats [1–15]; (ii) the rhizospheric soil is a good source for efficient isolation of calcifying bacteria as karst caves and secondary carbonate soils; iii) nutrient availability is a critical factor affecting microbial metabolism and precipitation activity [27]; (iv) the efficiency of the Biolog EcoPlate system as a tool for estimating calcification potentiality of environmental bacterial communities [60]; (v) the importance of bacterial calcification as a detoxification process (calcification by bacterial species may be an important phenotype for survival in high calcium environments as the studied soil—Table 1).

Table 5. Calcification parameters of the bacterial isolates.

	Timing [a]			Position [b]	Shape	Size	Amount [c]	Aggregates	Liquid Media
Strain	4 °C	20 °C	28 °C	28 °C	28 °C	28 °C	28 °C	28 °C	28 °C
P1	-	30	14	In/Out	G/O	M/L	+ + +	+	+
P2	-	13	8	In/Out	P/O	S/M	+ + + +	+ + +	+
P3	-	20	8	In/Out	G	M/L	+ + +	+ +	+ +
P4	-	23	10	In/Out	G	L	+ + +	+	-
P5	n.d.	n.d.	n.d.	In/Out	G/O	S	+	-	n.d.
M5 [d]	-	16	14	In/Out	G/O	S	+ + + +	++	-
P6	-	11	8	In	G/P	M/L	+ + + +	+ +	+++
P7	-	8	4	In/Out	G	L	++	+ +	-
P8	20	8	8	In/Out	G/P	S/L	++	+	++
P9	-	22	10	In/Out	G	M/L	+ + + +	++/+ + +	-
P10	-	13	8	In	G/P	S/M/L	++	-	+
P11	-	12	8	In/Out	G/P	S	+ + +	+/+ +	+

[a] Timing: Number of days required to start precipitation of CaCO$_3$ in B4 agar cultures at 4, 20, and 28 °C; [b] Position on B4 agarized medium of the deposited crystals (inside and/or outside the bacterial colony); [c] visually established by the microscope; [d] M5 is a mixed calcifying culture composed by the strains P12 (*Sphingomonas sanguinis*), P13 (not identified), and P14 (*Pseudomonas syringae*); n.d. = not determined; In = inside the colony; Out = outside the colony; In/Out = inside and outside the colonies; G = globular, spheric; O = oval; P = prismatic; S = small-sized; M = medium-sized (80–100 μm); L = large-sized (100–200 μm); CaCO$_3$ deposition: (-) = absent, (+) = trace, (+ +) = good, (+ + +) = very good, (+ + + +) = excellent.

Table 5 shows the calcification parameters of the bacterial isolates: Number of days required to start precipitation of $CaCO_3$ in B4 agar cultures at 4, 20, and 28 °C, position on B4 agarized medium of the deposited crystals, amount and size of the crystals visually established by the microscope, presence of bioliths aggregates and mineralization ability in liquid B4 medium.

At 28 °C, $CaCO_3$ formation took place rapidly and all the calcifying bacteria began to precipitate within two weeks (Table 5): High temperatures resulted in faster bacterial growth and thus, higher acetate consumption. Vice versa, at lower temperatures bacterial metabolism slows down and resulted, for instance, in slow rates of pH change and acetate consumption. In fact, when bacteria were cultivated at 20 °C, the time required for crystal precipitation was longer and all the calcifying isolates began to precipitate within four weeks (Table 5). Apart from strain P8, which precipitated after three weeks, bacterial growth and crystal deposition was not observed at 4 °C.

The microscopic observation showed that at 28 °C, $CaCO_3$ precipitation always takes place inside the bacterial colony and often in the immediate vicinity and/or in the culture medium, that is in a microenvironment conditioned by the direct presence of the cells. This means that for the deposition of $CaCO_3$ the presence of crystallization nuclei is not sufficient, but the role of microbial metabolism is necessary. No crystal formation was observed in uninoculated control plates. This confirms that $CaCO_3$ deposition was to be attributed to the bacterial cells.

Table 5 shows that, apart from strain P2 that deposited prismatic and oval crystals, all others strains deposited spherical crystals, as the only prevalent typology. About the size, 66.6% of the calcifying cultures deposited large-sized crystals (P1, P3, P4, P6, P7, P8, P9, and P10). Strains P2, P5, P11, and the mixed culture M5, precipitated crystals of smaller dimensions (Table 5). Regarding the amount of precipitated crystals, the most active cultures in calcification, were strains P1, P2, P3, P4, P6, P9, and P11 (Table 5). The remaining strains (P7, P8, and P10) showed quantitative characteristics of intermediate precipitation.

Strains with the larger amount of precipitates were also those that gave rise to more $CaCO_3$ aggregates (Table 5). Based on microscopic observation of the calcifying solid cultures, the mixed culture M5 produced larger crystal aggregates than the pure cultures of its constituent strains P12, P13, and P14. Actually, bacteria always live in bio-diversified community, metabolically interacting, thus extending their survival through co-metabolism. This is one of the reasons why it is sometimes preferable to use mixed calcifying cultures.

Strain P6, identified as *Corynebacterium urealyticum*, precipitates the largest amount of crystals in B4 liquid medium (Table 5). Strains P3 and P8, identified as *Clavibacter agropyri*, showed a good precipitation ability in liquid B4 medium. P1, P2, P10, and P11 showed very poor ability to precipitate in B4 medium; while P4, P7, P9, and the mixed culture M5 showed no precipitation activity in liquid medium (Table 5).

All the calcifying cultures showed carbonate-solubilization ability when grown on 0.14% $CaCO_3$, after four weeks of incubation at 28 °C (Table 6). *Vibrio tubiashii* (P2), *Clavibacter agropyri* (P3, P10), the not identified strains P1 (Figure 3), and P7 being particularly effective.

Apart from strain P9, identified as *Clavibacter agropyri*, which do not dissolve detectable amounts of carbonate, all the calcifying strains showed poor $CaCO_3$ solubilization ability, when grown on 2.5% $CaCO_3$ (Table 6). Three calcifying cultures (P1, P2, and M5) showed spotted solubilization (Table 6 and Figure 4). The mixed calcifying culture M5 showed solubilization activity in a lesser extent than expected.

Table 6. Solubilization activity of the calcifying bacterial cultures obtained from a rhizospheric soil (L'Aquila-Central Italy).

Strains	Identified Species	Solubilization	
		CaCO$_3$ (0.14%)	CaCO$_3$ (0.5%)
P1	Not identified	+ + +	+/- [a]
P2	*Vibrio tubiashii*	+ + +	+/-
P3	*Clavibacter agropyri*	+ + +	+/+ +
P4	*Clavibacter agropyri*	+ +	+ +
M5	*Sphingomonas sanguinis* (P12), Not identified (P13), *Pseudomonas syringae* (P14)	+ +	+
P6	*Corynebacterium urealyticum*	+ +	+ +
P7	Not identified	+ + +	+/+ +
P8	*Clavibacter agropyri*	+ +	+/-
P9	*Clavibacter agropyri*	+ +	-
P10	*Clavibacter agropyri*	+ + +	+
P11	*Sanguibacter suarezii*	+ +	+

[a] +/- = spotted solubilization.

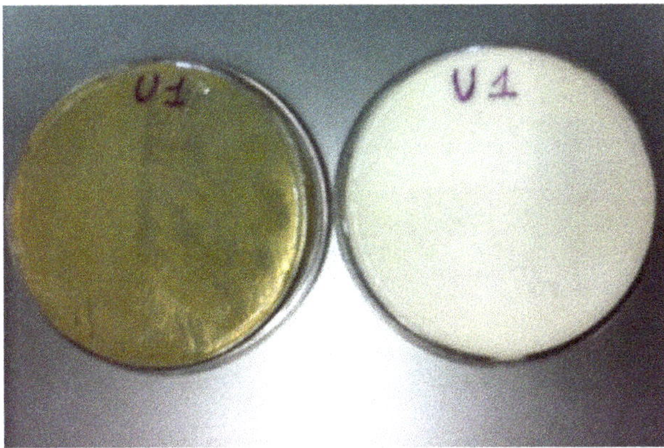

Figure 4. CaCO$_3$ solubilization, after 30 days incubation at 28 °C, by the not identified strain P1, firstly named U1. Note that the solubilization is almost complete at 0.14% (**left**), and as spots at 2.5% CaCO$_3$ (**right**).

3.5. Bacterial Calcification Activity by the Imagej Software

It is possible to evaluate the CaCO$_3$ precipitation efficiency of a calcifying strain taking into account different calcification parameters such as: (i) Time required to initiate precipitation, (ii) abundance of the precipitated crystals (which is related to the precipitation rate of the calcifying strain), (iii) crystal size, (iv) presence of crystal aggregates, and (v) CaCO$_3$ solubilization activity. The use of the ImageJ software allowed us to obtain a very effective BCP index, referred in this paper as "extent" (i.e., calcification level as percent average surface occupied by CaCO$_3$ crystals biodeposited on an agarized medium). The BCP extent, depending on all the calcification parameters mentioned above, allows to test in a single step the calcification aptitude of each bacterial strain, optimizing the screening program.

Figure 5 shows the percent average surface occupied by CaCO$_3$ crystals precipitated by the calcifying bacterial cultures, after 30 days of growth on B4 agar, at 28 °C.

Two groups of calcifying strains were identified on the basis of their calcification ability in terms of extent (%) of the biocemented area: A group of very highly active BCP bacteria, with percent average surface occupied by CaCO$_3$ crystals greater than 25% and group of less active bacteria with CaCO$_3$ percent average surface less than 25%. The calcifying strains belonging to the first group were strains

P2 (*Vibrio tubiashii*), P3, P4, and P9 (all three *Clavibacter agropyri*), P6 (*Corynebacterium urealyticum*), and P11 (*Sanguibacter suarezii*). All these bacterial strains produced in vitro CaCO$_3$ crystals abundant and medium/large sized, apart from strain P6 that formed smaller crystals (Table 5). They also showed a good aptitude to form large aggregates and vice versa a poor or absent CaCO$_3$ solubilization activity (Tables 5 and 6).

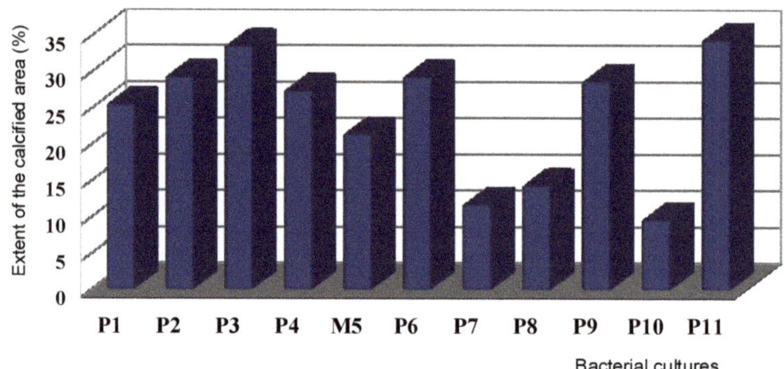

Figure 5. Percent average surface (BCP extent) occupied by the CaCO$_3$ crystals precipitated in vitro by the pure and mixed calcifying cultures, after 30 days of growth on B4 agar at 28 °C (the BCP-extent values were obtained through ImageJ software).

3.6. Calcifying Strain Selection for Sand Biocementation

Calcifying cultures for sand biocalcification experiments on Petri dish, were selected base on their morpho-physiological traits (Table 3) and calcification parameters (Tables 5 and 6, and Figure 5), as stated in Materials and Methods.

Among bacterial strains with BCP extent greater than 25% (Figure 5), we selected: (i) Strain P2, identified as *Vibrio tubiashii*, being the only one to have a not branched or irregular shape of the cell (Table 3). Short rods are certainly to be preferred compared to medium, long, or branched rods because of their greater possibility of penetration into the ground; (ii) strain P9, *Clavibacter agropyri*, for both the ability to growth in the absence of oxygen (Table 3) and its no solubilization activity of CaCO$_3$ at 2.5% (Table 6). Facultative anaerobes are preferred because of their tolerance for the restricted availability of oxygen that can be encountered in subsurface environments; this ability is fundamental because aerobic and anaerobic microzones coexist in the soil, since anaerobic conditions are abundant, particularly when the rain saturates the soil and gas exchange slows down almost completely [66]; and (iii) strain P11, identified as *Sanguibacter suarezii*, Gram-positive. Gram-positive bacteria are preferable to Gram-negative ones for their thicker wall, which makes them more resistant to changes in osmotic pressure, which is a typical condition for soil under consolidation or in recovery phase [66]. The mixed calcifying culture M5, belonging to the second group (Figure 5), was also included in the biocementation experiments.

After 20 days of biotreatment (at 28 °C, under static conditions) there was complete cementation of all the pre-treated sand samples (Figure 6).

No signs of calcification were observed on the sand sample not biotreated (Figure 6). This showed that the cementing action of the sand particles was to be attributed to the bacterial cells.

Strain P6 (*Corynebacterium urealyticum*), belonging to the first group of calcifying strains (Figure 5), was selected for deep sand biocementation base on its capability to precipitate CaCO$_3$ crystals in liquid B4 medium, to hydrolyze urea, and to grow in anaerobiosis (Table 5). The results obtained after one week (on the left) and two months (on the right) of treatment are shown in Figure 7.

Figure 6. Samples of biocemented sand through CaCO$_3$ precipitation by strains P2, P11, M5, and P9 (starting from the **upper left** to the **lower right**) after 20 days of incubation, at 28 °C. Untreated control at the **left bottom**. The plates were positioned vertically to highlight the biocementation of the sand, not present in the control.

Figure 7. Test for deep biocementation of 300 g of sea sand in Magenta vessel (60 × 60 and 50 mm height), by strain P6 (*Corynebacterium urealyticum*) grown on B4 liquid medium. On the **left** the sample after a week of treatment, on the **right** the sand-sample after two months of treatment (note the complete occlusion of visible pores by the biodeposition of CaCO$_3$ crystals).

After the biotreatment, the sand samples did not reach a cementation level sufficient to be subjected to the shear strength test, even in the presence of urea, calcium chloride, and low clay fraction [38,39]. Optimization of the deep biocementation experiments, fallowing outside the objectives of this work, has not been carried out.

3.7. SEM, XRD, and EDS Analyses

SEM observations were carried out to study the morphological characteristics of the biominerals precipitated and the involvement of bacteria in calcite biomineralization. A variety of crystal shapes were observed. The most frequent were single irregular spheres (Figure 8a,c–f), hemispheres (Figure 8a,e), and their irregular aggregates (Figure 8e, in particular). Most of the formed crystals had rough surfaces. SEM investigation also revealed the presence of bacterial cells in close contact with

CaCO$_3$ crystals (Figure 8a,c,d, and f, in particular). The presence of CaCO$_3$ associated with bacteria proves that bacteria served as nucleation sites during the mineralization process.

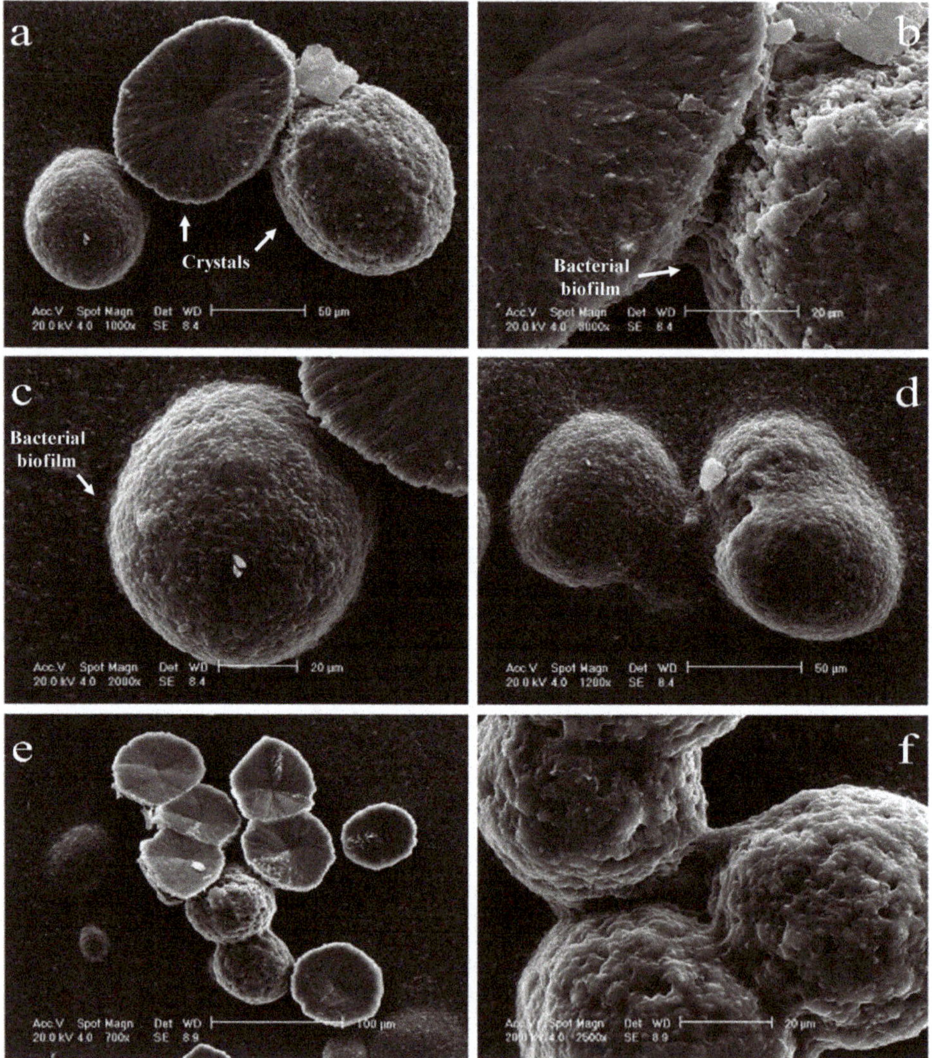

Figure 8. Scanning electron micrograph of CaCO$_3$ crystals and calcite aggregates deposited in vitro on agarized B4, at 28 °C by rhizospheric strains P1 (unidentified) (**a–d**) and P2 (*Vibrio tubiashii*) (**e,f**). (**a**) Bioliths, spherical and hemispherical; magnification 50 µm. (**b**) Enlargement of a, contact zone between two bioliths, during the aggregation phase, showing a bridge formed by biofilm and CaCO$_3$ crystals of neo-deposition; 20 µm magnification. (**c**) Enlargement of a, observe the significant amounts of bacterial cells in the background, 20 µm magnification. (**d**) Spherical bioliths completely immersed in the microbial biofilm, 50 µm magnification. (**e**) Aggregate formed by bioliths and hemi-bioliths, surrounded by bacterial cells; 100 µm magnification. (**f**) CaCO$_3$ aggregate of three bioliths, note the cementing action of the microbial biofilm; 20 µm magnification.

X-ray diffraction was carried out to identify the types of carbonate polymorph that were precipitated by the pure cultures selected for sand biocementation (P2 in Figure 9, P6, P9, and P11). In all the investigated cases, crystals precipitated by bacteria were both calcite, the rhombohedral form of $CaCO_3$ [R-3c] (card number 5–586), and vaterite, a hexagonal form of $CaCO_3$ [P63/mmc] (card number 24–30).

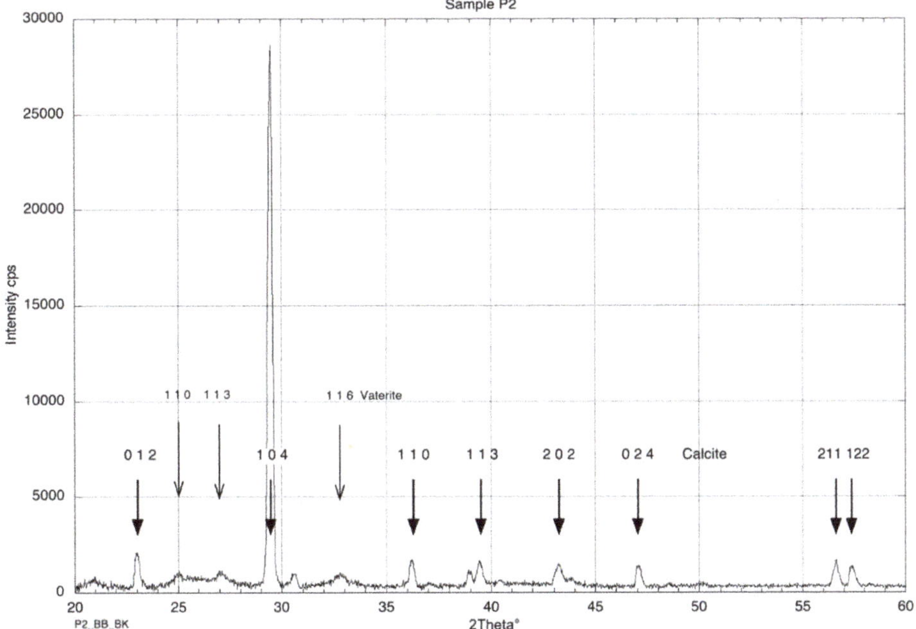

Figure 9. X-ray diffraction pattern from an agar dried culture of strain P2, *Vibrio tubiashii*. The spectrum shows the presence of a mixture of calcite and vaterite. The background level is due to the presence of the agar and of the adhesive tape.

Vaterite is the $CaCO_3$ polymorph with a high solubility that may be prone to dissolution by meteoric water and to a low degree of consolidation, by comparison to calcite. However, it was suggested that the microbially induced vaterite achieves similar degrees of stability as calcite by incorporating organic molecules [76].

SEM and EDS analyses of the sand samples biotreated on Petri dishes (Figure 6) showed that sand compaction was due to the cementing action of $CaCO_3$ bioliths that act as a bridge between the sand particles. The EDS spectra (Figure 10) showed Ca, C, O peaks that can be associated qualitatively with $CaCO_3$. The presence of high carbon, calcium, and oxygen peaks in the analyzed area, showed a high amount of $CaCO_3$. SEM image in Figure 10, shows that $CaCO_3$ crystals produced by strain P9, *Clavibacter agropyri*, coat and bridge the sand particles, cementing the matrix.

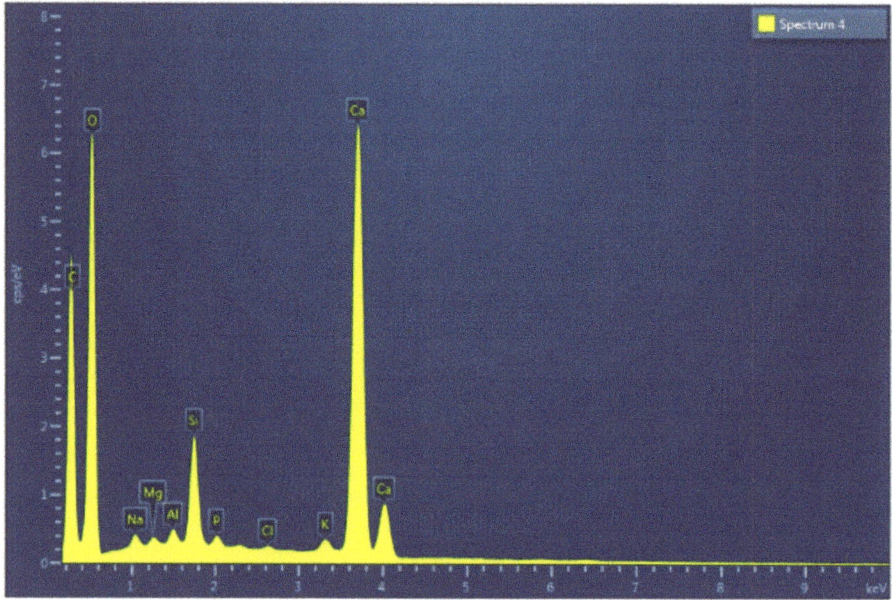

Figure 10. EDS spectrum and SEM image of the sand after 20 days of treatment at 28 °C with the strain P9 (*Clavibacter agropyri*).

According to EDS analysis, also the untreated sand presents, besides a high content of silica and oxygen, calcium and carbon but in low amount (Figure 11). As showed by the physico–chemical analyses (Table 2) and XRD (Figure 12), organic matter is the origin of C, while lithogenic carbonates could be the origin of Ca identified by EDS.

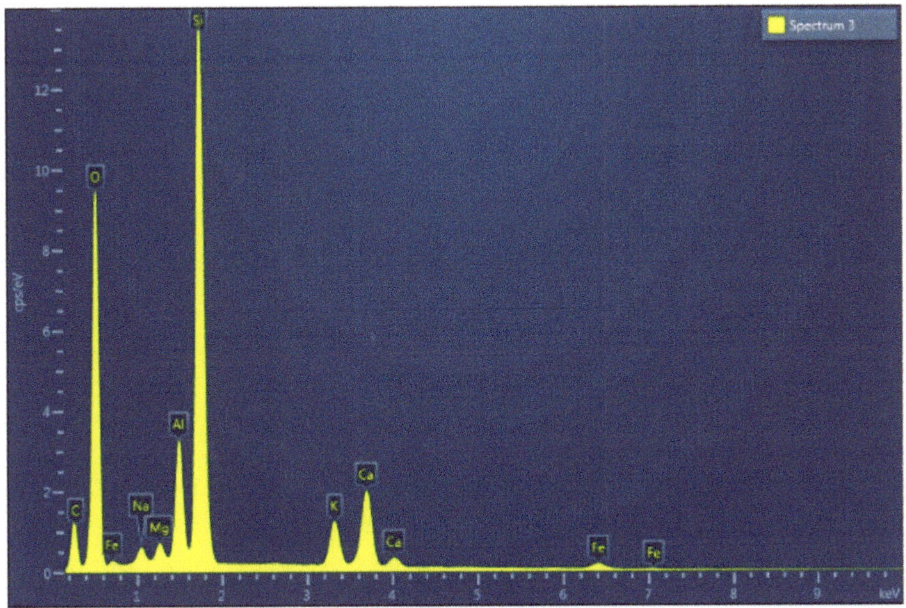

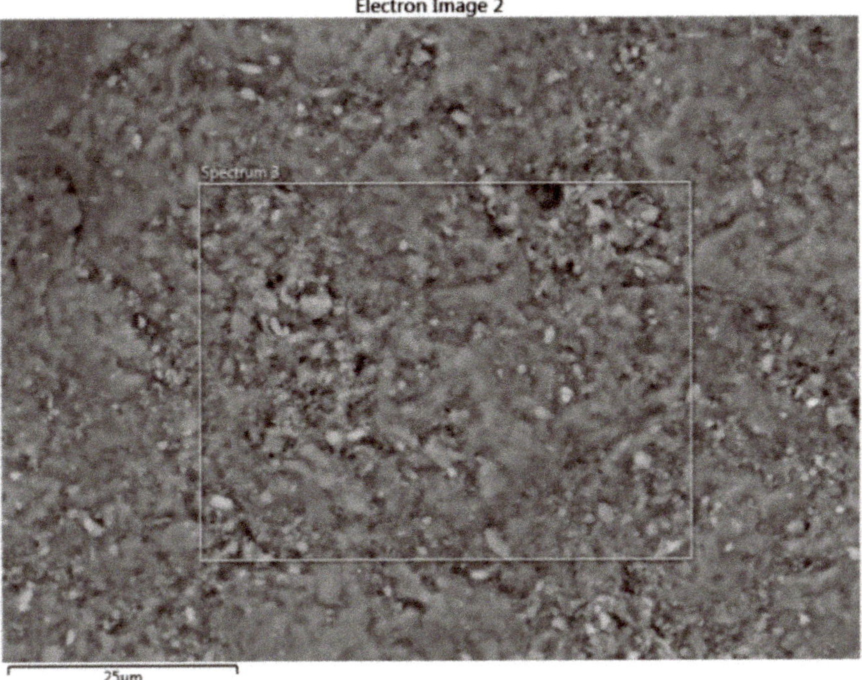

Figure 11. EDS spectrum and SEM image of the untreated sand.

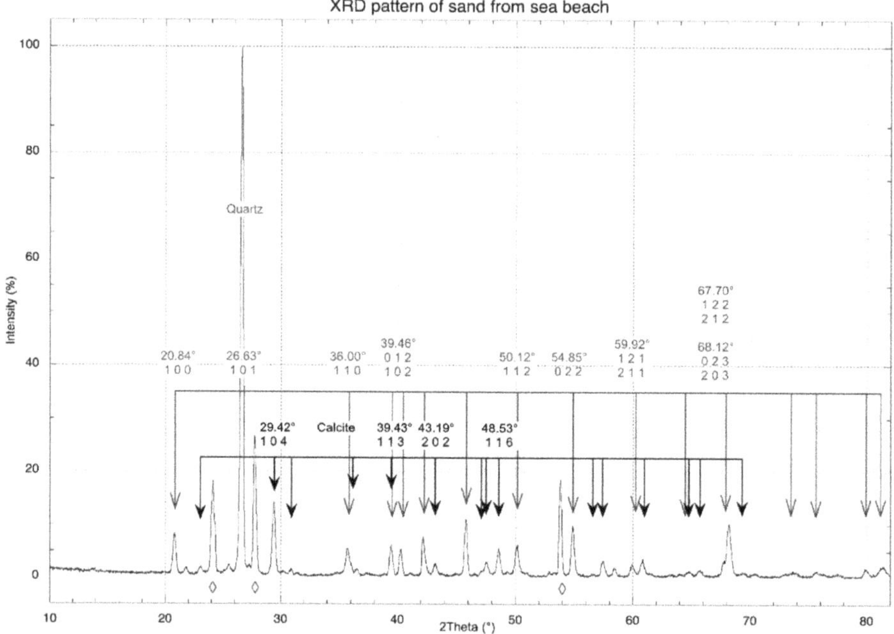

Figure 12. XRD pattern of the sea sand. The spectrum shows the prevalence of quartz and calcite.

4. Summary and Conclusions

Successful application of BCP in the different fields of interest, depends on a variety of factors. Much effort has been devoted to understand the factors governing BCP as well as the criteria for the selection of bacterial strains to optimize the biodeposition of $CaCO_3$ in a variety of biotechnological applications. To date, two strategies involving bacteria capable of inducing $CaCO_3$ precipitation have been mostly used: One in which an axenic culture is used directly for biotechnological purposes and another that utilizes the carbonatogenic potential of the autochthonous microbial communities. In this study, we adopted a novel approach based on the study (through the Biolog EcoPlates) of the BCP potential of the sampling site (a rhizospheric soil) and the use of selected (pure or mixed) calcifying cultures from this community. Until now, rhizospheric soil has been scarcely explored, with most of the studies focused on different calcium carbonate deposits in caves, freshwater, and seawater. Growth and biodiversity of the rhizospheric bacteria are highly influenced by type of soil, moisture content, temperature, chemicals, and above all by the plant species. Thus, choosing a proper rhizospheric soil (by knowing that microbial community dwelling the rhizospheric soil have the possibilities to perform ammonification of amino acids as well as mineralization of organic acids) it is possible to increase the screening efficiency for strong $CaCO_3$ depositor strains, likely due to increased number of bacteria useful to induce $CaCO_3$ precipitation. In this study, we used the B4 medium to isolate the cultivable heterotrophic calcifying microflora from the rhizospheric soil of *Pinus strobus*, obtaining a microbial consortium consisting exclusively of calcifying bacteria, although B4 was not a selective medium. This confirms how the combined use of Biolog EcoPlates (for a more targeted selection of the sampling site) and rhizospheric soil (as sampling site) can represent a useful tool to start a successful screening of potent calcifying cultures. To our knowledge this is the first study that shows the $CaCO_3$ biomineralization ability of *Vibrio tubiashii*, *Clavibacter agropyri* (*Corynebacterium*), *Corynebacterium urealyticum*, *Sanguibacter suarezii*, *Sphingomonas sanguinis*, and *Pseudomonas syringae*. As the metabolic response in the Biolog EcoPlates is at community level and involves in each well cooperative as well as competitive microbial behaviors, the real utility of the Biolog EcoPlates for a more targeted choice of the sampling site is conditioned by the use of mixed cultures whenever calcifying ability of the bacterial culture decreases or disappears following bacterial strain purification. Thus, it is important to preserve the BCP potential of the rhizospheric microbial community.

To select (pure or mixed) bacterial cultures suitable for sand biocementation, we performed an array of tests based on carbonate precipitation (i.e., $CaCO_3$ precipitation on solid media at different temperatures or in liquid condition at 28 °C), carbonate solubilization as well as physio-morphological analyses. To assess the extent of calcification by each bacterial strain, we applied ImageJ software as a new method for estimating BCP in solid media. Nevertheless, the positive results we obtained both in the enrichment and in the selection of bacteria useful for sand biocementation, the approach we proposed as well as the use of the ImageJ as a screening tool, need to be further investigated and optimized.

As a future goal, we are studying the role of cell interactions, namely the occurrence of a cell-to-cell communication system (quorum sensing), in the BCP process mediated by Gram-negative bacteria. Moreover, an optimization of the ImageJ method is in progress using a motorized microscope (Zeiss AXIO Imager.M2) with an Axiocam 503 camera, for the automatic choice of the visual fields and an accurate scale-bar determination and calibration.

Author Contributions: P.C. conceived the research question, contributed to methodology and data curation, wrote the draft paper, contributed to revision and editing; M.D.G. contributed to the analysis of the results, critical revision, and editing of the final manuscript.

Funding: This research received no external funding.

Acknowledgments: We thank L. Giansante for his assistance during the laboratory work, M. Giammatteo and L. Arrizza for assistance with SEM, M. Pellegrini and F. Matteucci for Biolog assistance and data elaboration. We also thank M. Passacantando and G. Cappuccio for collaboration in crystal XRD-analyses and graphic assistance.

Conflicts of Interest: The authors declare no conflict of interest.

References

1. Stocks-Fischer, S.; Galinat, J.K.; Bang, S.S. Microbiological precipitation of $CaCO_3$. *Soil Biol. Biochem.* **1999**, *31*, 1563–1571. [CrossRef]
2. Banks, E.D.; Taylor, N.M.; Gulley, J.; Lubbers, B.R.; Giarrizo, J.G.; Bullen, H.A.; Hoehler, T.M.; Barton, H.A. Bacterial calcium carbonate precipitation in cave environments: A function of calcium homeostasis. *Geomicrobiol. J.* **2010**, *27*, 444–454. [CrossRef]
3. Baskar, S.; Baskar, R.; Mauclaire, L.; Mckenzie, J.A. Role of microbial community in stalactite formation, Sahastradhara caves, Dehradun, India. *Curr. Sci.* **2005**, *88*, 1305–1308.
4. Boquet, E.; Boronat, A.; Ramos-Cormenzana, A. Production of calcite (calcium carbonate) crystals by soil bacteria is a general phenomenon. *Nature* **1973**, *246*, 527–529. [CrossRef]
5. Braissant, O.; Cailleau, G.; Dupraz, C.; Verrecchia, E.P. Bacterially induced mineralization of calcium carbonate in terrestrial environments: The role of exopolysaccharides and amino acids. *J. Sediment. Res.* **2003**, *73*, 485–490. [CrossRef]
6. Cacchio, P.; Ercole, C.; Lepidi, A. Calcium carbonate deposition in limestone caves: Microbiological aspects. *Subterr. Biol.* **2003**, *1*, 57–63.
7. Cacchio, P.; Ercole, C.; Cappuccio, G.; Lepidi, A. Calcium carbonate precipitation by bacterial strains isolated from a limestone cave and from a loamy soil. *Geomicrobiol. J.* **2003**, *20*, 85–98. [CrossRef]
8. Cacchio, P.; Contento, R.; Ercole, C.; Cappuccio, G.; Preite Martinez, M.; Lepidi, A. Involvement of microorganisms in the formation of carbonate speleothems in the Cervo Cave (L'Aquila-Italy). *Geomicrobiol. J.* **2004**, *21*, 497–509. [CrossRef]
9. Cacchio, P.; Ercole, C.; Lepidi, A. Evidences for bioprecipitation of pedogenic calcite by calcifying bacteria from three different soils of L'Aquila Basin (Abruzzi, Central Italy). *Geomicrobiol. J.* **2015**, *32*, 701–711. [CrossRef]
10. Danielli, H.M.C.; Edington, M.A. Bacterial calcification in limestone caves. *Geomicrobiology* **1983**, *3*, 1–16. [CrossRef]

11. Lipman, C.B. Further studies on marine bacteria with special reference to the drew hypothesis on CaCO$_3$ precipitation on sea. *Carnegie Inst.* **1929**, *39*, 231–248.

12. Morita, R.Y. Calcite precipitation by marine bacteria. *Geomicrobiology* **1980**, *2*, 63–82. [CrossRef]

13. Northup, D.E.; Reysenbach, A.L.; Pace, N.R. Microorganisms and speleonthems. In *Cave Minerals of the World*, 2nd ed.; Hill, C., Forti, P., Eds.; National Speleological Society: Huntsville, AL, USA, 1997; pp. 261–266. ISBN 1-879961-07-5.

14. Novitsky, J.A. Calcium carbonate precipitation by marine bacteria. *Geomicrobiology* **1981**, *2*, 375–388. [CrossRef]

15. Riding, R. Microbial carbonates: The geological record of calcified bacterial-algal mats and biofilms. *Sedimentology* **2000**, *47*, 179–214. [CrossRef]

16. Zhuang, D.; Yan, H.; Tuckerd, M.E.; Zhao, H.; Han, Z.; Zhao, Y.; Sun, B.; Li, D.; Pan, J.; Zhao, Y.; et al. Calcite precipitation induced by Bacillus cereus MRR2 cultured at different Ca^{2+} concentrations: Further insights into biotic and abiotic calcite. *Chem. Geol.* **2018**, *500*, 64–87. [CrossRef]

17. Castanier, S.; Le Metayer-Levrel, G.; Perthuisot, J.-P. Ca-carbonates precipitation and limestone genesis—The microbiogeologist point of view. *Sediment. Geol.* **1999**, *126*, 9–23. [CrossRef]

18. Hammes, F.; Verstraete, W. Key roles of pH and calcium metabolism in microbial carbonate precipitation. *Rev. Environ. Sci. Biotechnol.* **2002**, *1*, 3–7. [CrossRef]

19. Douglas, S.; Beveridge, T.J. Mineral formation by bacteria in natural microbial communities. *FEMS Microbiol. Ecol.* **1998**, *26*, 79–88. [CrossRef]

20. Castanier, S.; Métayer-Levrel, G.L.; Perthuisot, J.-P. Bacterial Roles in the Precipitation of Carbonate Minerals. In *Microbial Sediments*; Riding, R.E., Awramik, S.M., Eds.; Springer: Heidelberg/Berlin, Germany, 2000; pp. 32–39. [CrossRef]

21. Schultze-Lam, S.; Fortin, D.; Davis, B.S.; Beveridge, T.J. Mineralisation of bacterial surfaces. *Chem. Geol.* **1996**, *132*, 171–181. [CrossRef]

22. Rivadeneyra, M.A.; Delgado, G.; Ramos-Cormenzana, A.; Delgado, R. Biomineralization of carbonates by Halomonas eurihalina in solid and liquid media with different salinities: Crystal formation sequence. *Res. Microbiol.* **1998**, *149*, 277–287. [CrossRef]

23. Southam, G. Bacterial Surface-Mediated Mineral Formation. In *Environmental Microbe-Metal Interactions*; Lovley, D., Ed.; ASM Press: Washington, DC, USA, 2000; pp. 257–276.

24. Braissant, O.; Decho, A.W.; Dupraz, C.; Glunk, C.; Przekop, K.M.; Visscher, P.T. Exopolymeric substances of sulfate-reducing bacteria: Interactions with calcium at alkaline pH and implication for formation of carbonate minerals. *Geobiology* **2007**, *5*, 401–411. [CrossRef]

25. Decho, A.W. Overview of biopolymer-induced mineralization: What goes on in biofilms? *Ecol. Eng.* **2009**, *30*, 1–8. [CrossRef]

26. Ercole, C.; Cacchio, P.; Botta, A.L.; Centi, V.; Lepidi, A. Bacterially induced mineralization of calcium carbonate: The role of exopolysaccharides and capsular polysaccharides. *Microsc. Microanal.* **2007**, *13*, 42–50. [CrossRef] [PubMed]

27. Braissant, O.; Verrecchia, E.P.; Aragno, M. Is the contribution of bacteria to terrestrial carbon budget greatly underestimated? *Naturwissenschaften* **2002**, *89*, 366–370. [CrossRef] [PubMed]

28. Arp, G.; Volker, T.; Reimer, A.; Michaelis, W.; Reitner, J. Biofilm exopolymers control microbialite formation at thermal springs discharging into the alkaline Pyramid Lake, Nevada, USA. *Sediment. Geol.* **1999**, *126*, 159–176. [CrossRef]

29. McConnaughey, T.A.; Whelan, J.F. Calcification generates protons for nutrient and bicarbonate uptake. *Earth Sci. Rev.* **1997**, *42*, 95–117. [CrossRef]

30. Seifan, M.; Samani, A.K.; Berenjian, A. Induced calcium carbonate precipitation using Bacillus species. *Appl. Microbiol. Biotechnol.* **2016**, *100*, 9895–9906. [CrossRef]

31. Wright, D.T. The role of sulphate-reducing bacteria and cyanobacteria in dolomite formation in distal ephemeral lakes of Coorong region, South Australia. *Sed. Geol.* **1999**, *126*, 147–157. [CrossRef]

32. Arp, G.; Reimer, A.; Reitner, J. Photosynthesis-induced biofilm calcification and calcium concentrations in Phanerozoic oceans. *Science* **2001**, *292*, 1701–1704. [CrossRef]

33. Dittrich, M.; Obst, M. Are picoplankton responsible for calcite precipitation in lakes? *Ambio* **2004**, *33*, 559–564. [CrossRef]

34. Plée, K.; Pacton, M.; Ariztegui, D. Discriminating the role of photosynthetic and heterotrophic microbes triggering low-Mg calcite precipitation in freshwater biofilms (Lake Geneva, Switzerland). *Geomicrobiol. J.* **2010**, *27*, 391–399. [CrossRef]

35. Power, I.M.; Wilson, S.A.; Small, D.P.; Dipple, G.M.; Wan, W.K.; Southam, G. Microbially mediated mineral carbonation: Roles of phototrophy and heterotrophy. *Environ. Sci. Technol.* **2011**, *45*, 9061–9068. [CrossRef] [PubMed]

36. Ehrlich, H.L. Geomicrobiology: Its significance for geology. *Earth-Sci. Rev.* **1998**, *45*, 45–60. [CrossRef]

37. Zhu, T.; Dittrich, M. Carbonate precipitation through microbial activities in natural environment, and their potential in biotechnology: A review. *Front. Bioeng. Biotechnol.* **2016**, *4*, 4. [CrossRef] [PubMed]

38. Cardoso, R.; Pires, I.; Duarte, S.O.D.; Monteiro, G.A. Effects of clay's chemical interactions on biocementation. *Appl. Clay Sci.* **2018**, *156*, 96–103. [CrossRef]

39. Morales, L.; Garzón, E.; Sánchez-Soto, S.P.J.; Romero, E. Microbiological induced carbonate (CaCO$_3$) precipitation using clay phyllites to replace chemical stabilizers (cement or lime). *Appl. Clay Sci.* **2019**, *174*, 15–28. [CrossRef]

40. Anbu, P.; Kang, C.H.; Shin, Y.J.; So, J.S. Formation of calcium carbonate minerals by bacteria and its multiple applications. *SpringerPlus* **2016**, *5*, 250. [CrossRef]

41. Vekariya, M.S.; Pitroda, J. Concrete: New Era for Construction Industry. *Intern. J. Eng. Tre. Tech.* **2013**, *4*, 4128–4137.

42. Chahal, N.; Rajor, A.; Siddique, R. Calcium carbonate precipitation by different bacterial strains. *Afr. J. Biotechnol.* **2011**, *10*, 8359–8372.

43. Garland, J.L.; Mills, A.L. Classification and characterization of heterotrophic microbial communities on the basis of patterns of community-level sole-carbon-source utilization. *Appl. Environ. Microbiol.* **1991**, *57*, 2351–2359.

44. Dakora, F.D.; Phillips, D.A. Root exudates as mediators of mineral acquisition in low-nutrient environments. *Plant and Soil* **2002**, *245*, 35–47. [CrossRef]

45. Abràmoff, M.D.; Magelhaes, P.J.; Ram, S.J. Image processing with ImageJ. *Biophotonics Inter.* **2004**, *11*, 36–42.

46. Schneider, C.A.; Rasband, W.S.; Eliceiri, K.W. NIH Image to ImageJ: 25 years of image analysis. *Nat. Methods* **2012**, *9*, 671–675. [CrossRef] [PubMed]

47. Garland, J.L. Analytical approaches to the characterization of samples of microbial communities using patterns of potential C source utilization. *Soil Biol. Biochem.* **1996**, *28*, 213–222. [CrossRef]

48. Frąc, M.; Oszust, K.; Lipiec, J. Community level physiological profiles (CLPP), characterization and microbial activity of soil amended with dairy sewage sludge. *Sensors* **2012**, *12*, 3253–3268. [CrossRef]

49. Zak, J.C.; Willig, M.R.; Moorhead, D.L.; Wildman, H.G. Functional diversity of microbial communities: A quantitative approach. *Soil Biol. Biochem.* **1994**, *26*, 1101–1108. [CrossRef]

50. Supraja, N.; Prasad, T.N.V.K.V. Extracellular synthesis of Calcium Precipitating Bacteria Isolated from Rock Environment Region and its Characterization Studies. *Adv. Bioequiv. Bioavailab.* **2018**, *1*. [CrossRef]

51. Otlewska, A.; Gutarowska, B. Environmental parameters conditioning microbially induced mineralization under the experimental model conditions. *Acta Biochim. Pol.* **2016**, *63*, 343–351. [CrossRef]

52. Marvasi, M.; Visscher, P.T.; Perito, B.; Mastromei, G.; Martinez, L. Physiological requirements for carbonate precipitation during biofilm development of Bacillus subtilis etfA mutant. *FEMS Microbiol. Ecol.* **2010**, *71*, 341–350. [CrossRef]

53. Urzì, C.; Garcia-Valles, M.; Vendrell, M.; Pernice, A. Biomineralization process on rock and monument surfaces observed in field and in laboratory conditions. *Geomicrobiol. J.* **1999**, *16*, 39–54.

54. Baskar, S.; Baskar, R.; Mauclaire, L.; McKenzie, J.A. Microbially induced calcite precipitation in culture experiments: Possible origin for stalactites in Sahastradhara caves, Dehradun, India. *Curr. Sci.* **2006**, *90*, 58–64.

55. Smibert, R.M.; Krieg, N.R. General characterization. In *Manual Methods for General Bacteriology*; Gerhardt, P., Ed.; American Society for Microbiology: Washington, DC, USA, 1981; pp. 409–443.

56. Christensen, W.B. Urea Decomposition as a Means of Differentiating Proteus and Paracolon Cultures from Each Other and from Salmonella and Shigella Types. *J. Bacteriol.* **1946**, *52*, 461–466. [PubMed]

57. MacFaddin, J.F. *Biochemical Tests for Identification of Medical Bacteria*, 3rd ed.; Williams & Wilkins: Philadelphia, PA, USA, 2000; p. 298. ISBN 0683053183.

58. NORMAL Commission. NORMAL 9/88 recommendations. In *Autotrophic and Heterotrophic Microflora: Culture Methods for Isolation*; Consiglio Nazionale delle Ricerche/Istituto Centrale per il Restauro: Rome, Italy, 1990.

59. Martino, T.; Salamone, P.; Zagari, M.; Urzì, C. Adesione a substrati solidi e solubilizzazione del CaCO3 quale misura della capacità deteriorante di batteri isolati dal marmo Pentelico. In Proceedings of the XI Meeting of the Italian Society of General Microbiology and Microbial Biotechnology (SIMGBM), Gubbio, Italy, 25–28 September 1992; pp. 249–250.

60. Andrei, A.-S.; Păuşan, M.R.; Tămaş, T.; Har, N.; Barbu-Tudoran, L.; Leopold, N.; Banciu, H.L. Diversity and biomineralization potential of the epilithic bacterial communities inhabiting the oldest public stone monument of Cluj-Napoca (Transylvania, Romania). *Front. Microbiol.* **2017**, *8*, 372. [CrossRef] [PubMed]

61. Jones, D.L. Organic acids in the rhizosphere—A critical review. *Plant Soil* **1998**, *205*, 25–44. [CrossRef]

62. Florenzano, G. *Fondamenti di Microbiologia del Terreno*; REDA: Rome, Italy, 1983; p. 123.

63. Marilley, J.; Aragno, M. Phylogenetic diversity of bacterial communities differing in degree of proximity of Lolium perenne and Trifolium repens roots. *Appl. Soil Ecol.* **1999**, *13*, 127–136. [CrossRef]

64. Atlas, R.M.; Bartha, R. *Microbial Ecology: Fundamentals and Applications*, 4th ed.; The Benjamin/Cummings Publishing Company: San Francisco, CA, USA, 1998; ISBN 0805306552.

65. Kourtev, P.S.; Ehrenfeld, J.G.; Haggblom, M. Exotic plant species alter the microbial community structure and function in the soil. *Ecology* **2002**, *83*, 3152–3166. [CrossRef]

66. Ivanov, V.; Chu, J. Applications of microorganisms to geotechnical engineering for bioclogging and biocementation of soil in situ. *Rev. Environ. Sci. Biotechnol.* **2008**, *7*, 139–153. [CrossRef]

67. Burbank, M.B.; Weaver, T.J.; Williams, B.C.; Crawford, R.L. Urease activity of ureolytic bacteria isolated from six soils in which calcite was precipitated by indigenous bacteria. *Geomicrobiol. J.* **2012**, *29*, 389–395. [CrossRef]

68. Davis, M.J.; Gillaspie, A.G.; Vidaver, A.K.; Harris, R.W. Clavibacter: A new genus containing some phytopathogenic coryneform bacteria, including, Clavibacter xyli subsp. xyli sp. nov., subsp. nov. and Clavibacter xyli subsp. cynodontis subsp. nov., pathogens that cause ratoon stunting disease of sugarcane and bermudagrass stunting disease. *Int. J. Bacteriol.* **1984**, *34*, 107–117.

69. Riley, I.T. Serological relationships between strains of coryneform bacteria responsible for annual ryegrass toxicity and other plant pathogenic corynebacteria. *Int. J. Syst. Bacteriol.* **1987**, *37*, 153–159. [CrossRef]

70. Riley, I.T.; Reardon, T.B.; McKay, A.C. Genetic Analysis of plant pathogenic bacteria in the genus clavibacter using allozyme electrophoresis. *J. Gen. Microbiol.* **1988**, *134*, 3025–3030. [CrossRef]

71. Fernandez-Garayzabal, J.F.; Dominguez, L.; Pascual, C.; Jones, D.; Collins, M. Phenotypic and phylogenetic characterization of some unknown coryneform bacteria isolated from bovine blood and milk: Description of Sanguibacter gen. nov. *Lett. Appl. Microbiol.* **1995**, *20*, 69–75. [CrossRef] [PubMed]

72. Thompson, F.L.; Iida, T.; Swings, J. Biodiversity of *Vibrios*. *J. Microbiol. Mol. Biol. Rev.* **2004**, *68*, 403–431. [CrossRef] [PubMed]

73. Tubiash, H.S.; Chanley, P.E.; Leifson, E. Bacillary necrosis, a disease of larval and juvenile bivalve mollusks. I. Etiology and epizootiology. *J. Bacteriol.* **1965**, *90*, 1036–1044. [PubMed]

74. Okada, K.; Iida, T.; Kita-Tsukamoto, K.; Takeshi, H. Vibrios commonly possess two chromosomes. *J. Bacteriol.* **2005**, *187*, 752–757. [CrossRef] [PubMed]

75. Rivadeneyra, M.A.; Delgado, R.; del Moral, A.; Ferrer, M.R.; Ramos-Cormenzana, A. Precipitation of calcium carbonate by Vibrio spp. from an inland saltern. *FEMS Microbiol. Ecol.* **1994**, *13*, 197–204. [CrossRef]

76. Rodriguez-Navarro, C.; Jimenez-Lopez, C.; Rodriguez-Navarro, A.; Gonzalez-Muñoz, M.T.; Rodriguez-Gallego, M. Complex biomineralized vaterite structures encapsulating bacterial cells. *Geochim. Cosmochim. Acta* **2007**, *71*, 1197–1213. [CrossRef]

Article

Fossilized Endolithic Microorganisms in Pillow Lavas from the Troodos Ophiolite, Cyprus

Diana-Thean Carlsson [1,2], Magnus Ivarsson [1,3,]* and Anna Neubeck [4,]*

[1]   Department of Paleobiology, Swedish Museum of Natural History, 114 18 Stockholm, Sweden; diana.carlsson@nrm.se
[2]   Department of Earth Sciences, University of Hamburg, 20146 Hamburg, Germany
[3]   Department of Biology, University of Southern Denmark, 5230 Odense, Denmark
[4]   Department of Earth Sciences, Palaeobiology, Uppsala University, 752 36 Uppsala, Sweden
*   Correspondence: magnus.ivarsson@nrm.se (M.I.); anna.neubeck@geo.uu.se (A.N.)

Received: 23 August 2019; Accepted: 18 October 2019; Published: 23 October 2019

Abstract: The last decade has revealed the igneous oceanic crust to host a more abundant and diverse biota than previously expected. These underexplored rock-hosted deep ecosystems dominated Earth's biosphere prior to plants colonized land in the Ordovician, thus the fossil record of deep endoliths holds invaluable clues to early life and the work to decrypt them needs to be intensified. Here, we present fossilized microorganisms found in open and sealed pore spaces in pillow lavas from the Troodos Ophiolite (91 Ma) on Cyprus. A fungal interpretation is inferred upon the microorganisms based on characteristic morphological features. Geochemical conditions are reconstructed using data from mineralogy, fluid inclusions and the fossils themselves. Mineralogy indicates at least three hydrothermal events and a continuous increase of temperature and pH. Precipitation of 1) celadonite and saponite together with the microbial introduction was followed by 2) Na and Ca zeolites resulting in clay adherence on the microorganisms as protection, and finally 3) Ca carbonates resulted in final fossilization and preservation of the organisms in-situ. Deciphering the fossil record of the deep subseafloor biosphere is a challenging task, but when successful, can unlock doors to life's cryptic past.

Keywords: deep biosphere; fossilized microorganisms; Ophiolite

---

1. Introduction

The igneous oceanic crust has been put forward as the largest potential microbial habitat on Earth [1,2]. However, due to difficulties in sampling live species, this vast biosphere is largely unexplored. Additional to a few successful molecular studies [3–6], paleontological material has been used to study life in the oceanic crust [7,8]. Fossilized microorganisms holds paleobiological information on organismal affinity [9–11], microbe–mineral interactions [12], and clues to early evolution of multicellularity [13]. Microbial life in terrestrial and marine crust is usually referred to as the deep biosphere, which today represents between one-tenth to one-third of all living biomass on Earth while a majority of the rest are land plants [14]. However, before vegetation colonized land (~400 Ma) the deep biosphere was the predominant reservoir for living biomass [14]. Elucidating the fossil record of deep endolithic life is thus crucial to understand life's history on Earth. For example how and when eukaryotes evolved to occupy crucial habitats and how this process influenced the eventual dominance of the total biosphere by multicellular eukaryotes. Investigations of subseafloor paleontological material has further displayed the geobiological role deep microorganisms play, including microbe–mineral interactions like weathering of secondary carbonates and zeolites, or formation of Fe-, Mn-oxides and clays [8,10,12,15]. The connection between microorganisms, element cycling, and mineralization in these deep niches are essential for the preservation and fossilization of microorganisms but far from understood [8].

The oceanic crust is a heterogeneous environment influenced by plate tectonics, volcanism, sediment overburden, but also characterized by more local anomalies like hydrothermal activity and methane seeps. Recycling and renewal of oceanic crust occur along spreading centers, subduction zones, and seamounts, and it is mainly these areas that are sampled and studied with geomicrobiology in mind [2]. However, the uncertainty in environmental, geological, and geochemical parameters that drives and support deep life are still not well understood. Further investigations are needed to understand the connection between microorganisms and their surrounding geological environment.

This study focus on veins and vesicles in pillow lavas from the 91 Ma Troodos ophiolite on Cyprus [16]. Previous studies has shown the Troodos pillow lavas to host ichnofossils in volcanic glass [17,18] but the current study focus on filamentous body fossils. To understand the past living habitat of the microorganisms, the associated mineralogy and fluid inclusions were investigated. Results indicate a direct relationship between microbial colonization, associated mineralogy, hydrothermal fluids, and fossilization.

Geological Setting and Sampled Localities

Cyprus is a part of the Anatolian plate in the Mediterranean Sea. It developed in a supra-subduction zone as an ocean spreading ridge in the Tethys Ocean during the Upper Cretaceous [19]. Crystallization and formation of the ophiolite occurred below the carbonate compensation depth. Active spreading ceased during the mid-Miocene when the oceanic crust was pushed up onto the African plate, creating the Marmonian and Troodos Terrains [20]. Continued movement during the upper Miocene, Pliocene, and Pleistocene contributed to accreted sedimentary sequences, the Circum Troodos Sedimentary and the Keryneia Terrain.

We investigated samples from the Troodos Terrain and the Troodos ophiolite therein. Our focus is on fresh basalt in pillow lavas, far from glassy cooling rims, commonly found in association with this type of setting. The sampling was done in Mathiatis mine, an old open pit in which Au, Ag and Fe was mined up until operations ceased in 1987. Two samples (A and B) were sampled at coordinates Latitude 34 58′31.7, Longitude 33 20′49.3 and Latitude 34 58′31.9, Longitude 33 20′47.0, respectively (Figure 1). The mine stratigraphy involves pyrite-rich quartzite at the bottom of the mine overlain by brecciated and chloritized basalts, followed by a younger lava flow. Sampling for the current study was done on the basalt of the lower pillow lavas at the top. For this study only samples from the pillow interiors were used to exclude later weathering.

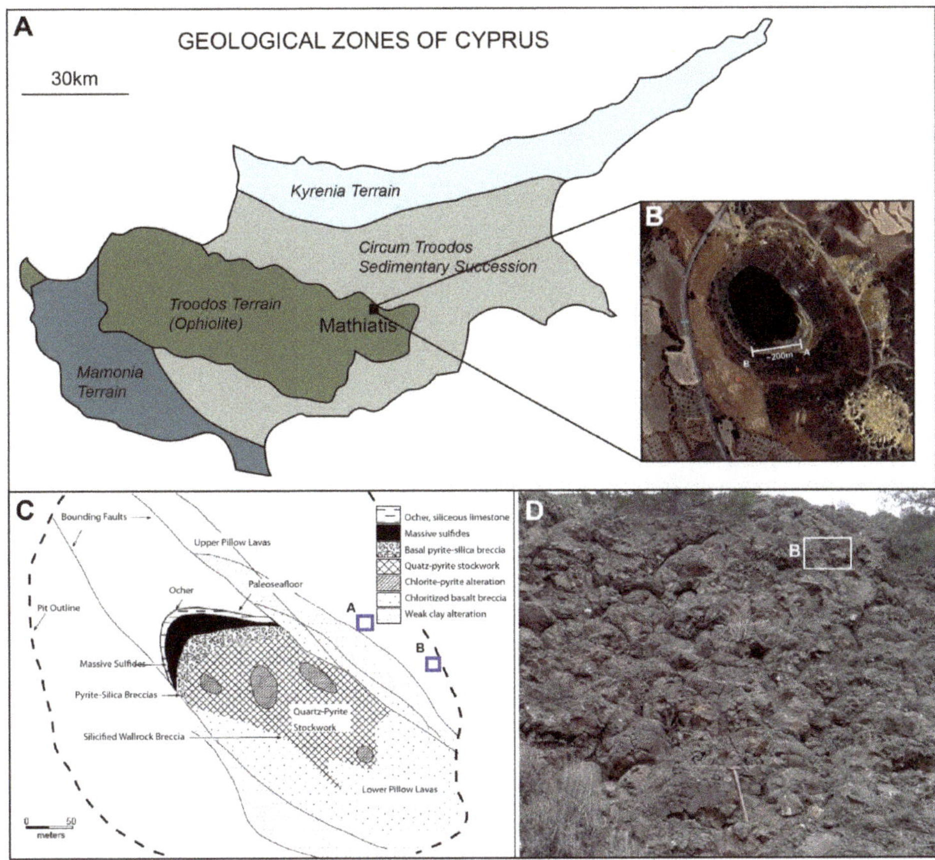

Figure 1. (**A**) Map of the geological zones of Cyprus. (**B**) Insert showing sampling locations 07A and 07B. From Google maps. (**C**) Bedrock map of the Mathiatis mine with sampling locations. (**D**) Pillow lavas at the Mathiatis mine with sampling point 07B. Geological hammer for scale.

2. Materials and Methods

2.1. Microscopy

Thin sections were prepared by Vancouver Petrographics Limited and ABC Ahead in Poland with a thickness of 150–200 µm according to protocols by [21]. Main mineralogy, secondary mineralization, and opaque minerals were analyzed in thin sections using a Nikon Optiphot2 polarization microscope fitted with a Las Ez3 camera. Putative fossilized microorganisms were determined by using a Nikon SMZ1500 microscope equipped with a Nikon D80 camera. Optical microscopy was carried out at the Department of Geological Sciences, Stockholm University, and the Department of Palaeobiology at the Swedish Museum of Natural History in Stockholm.

2.2. Environmental Scanning Electron Microscopy

Environmental Scanning Electron microscopy (ESEM) was carried out at the Department of Geological Science at Stockholm University. For this study, a Philips XL-30-ESEM-FEG650 and energy-dispersive spectroscopy (EDS) were used. The electron beam was set to 20 kV and the probe current to 6.00 nA at a working distance of 10 mm. Low vacuum was applied to the sample chamber, giving rise to a conductive atmosphere layer such that carbon coating was not needed for analyses.

2.3. Raman Spectroscopy

Raman spectra were collected at the Department of Geological Sciences at Stockholm University, using a confocal laser Raman microspectrometer (Horiba instrument LabRAM HR 800; Horiba Jobin Yvon, Villeneuve d'Ascq, France), equipped with a multichannel air-cooled (−70 °C) 1024 × 256 pixel CCD (charge-coupled device) detector. Spectra were obtained with 1800 lines/mm grating. Excitation was provided by an Ar-ion laser (λ = 514 nm) source. Spectra were recorded using a low laser power of 0.1–1 mW at the sample surface to avoid laser-induced alteration of the samples. Analyses were carried out using an Olympus BX41 microscope coupled to the instrument, and the laser beam was focused through 80× (hand specimens, working distance of 8 mm) and 100× (thin sections) objectives to obtain a spot size of about 1 μm. The spectral resolution was 0.3 cm^{-1}/pixel, with a typical exposure time of 10 s and with 10 accumulations. The accuracy of the instrument was controlled by repeated use of a silicon wafer calibration standard with a characteristic Raman line at 520.7 cm^{-1}. The Raman spectra were achieved with LabSpec 5 software. Minerals in thin sections and hand specimens of collected samples were identified by Laser Raman spectroscopy and comparisons with reference spectra in the RRUFF database [22].

2.4. Microthermometry

Microthermometric analyses on fluid inclusions in calcite were performed with a Linkam THM 600 stage mounted on a Nikon microscope utilizing a 40× long-working-distance objective. The working range of the stage is −196 °C to +600 °C [23]. Calibration was made using SynFlinc®synthetic fluid inclusions and well-defined natural inclusions in Alpine quartz. The reproducibility was ±0.1 °C for temperatures below 40 °C and ±0.5 °C for temperatures above 40 °C.

3. Results

3.1. Mineralogy and Filamentous Structures

The mineralogy of the veins and vesicles was dominated by calcite, zeolites (analcime, mordenite), clays (illite, smectite, montmorillonite) and iron oxides (goethite) according to Raman spectroscopy, see results below. Filaments were found in both open (Figure 2a) and carbonate- and zeolite filled (Figure 2b) veins and vesicles. The latter was studied by thin sections while the first just as hand specimens. They all originate from a clay/iron oxide film lining the interior of the pore space (Figure 3a,b) and protrude from the walls of the host rock into the pore space. The filaments occured either as single features or in abundance, forming complex networks (Figure 2a). They were curvi-linear (Figure 2a), frequently branched (Figure 3c), and occasionally partitioned by septa-like walls (Figure 2b insets). The septation was mostly repetitive and divide the filaments in individual compartments, and in a few filaments a nucleus occured in the center of each compartment (Figure 2b insets). The diameter of the filaments ranged from 5 to 50 μm, and the length ranged from ~50 to several hundred μms. In most cases, the length were reduced by thin section preparation or by damage in open pore spaces. Many filaments also had a central strand that ranged from a few to 15 μm in diameter, making up between a few percent to more than half the diameter of the entire filaments (Figure 3d). The central strand was distinct by its darker brown-red coloration compared to the greyish margins (Figure 3d), and it was possible to see branching of the central strand following the general branching of the filaments (Figure 3e). Occasionally, filaments in open pore space had a terminal precipitation of zeolites forming a swelling at the filament tip (Figure 4).

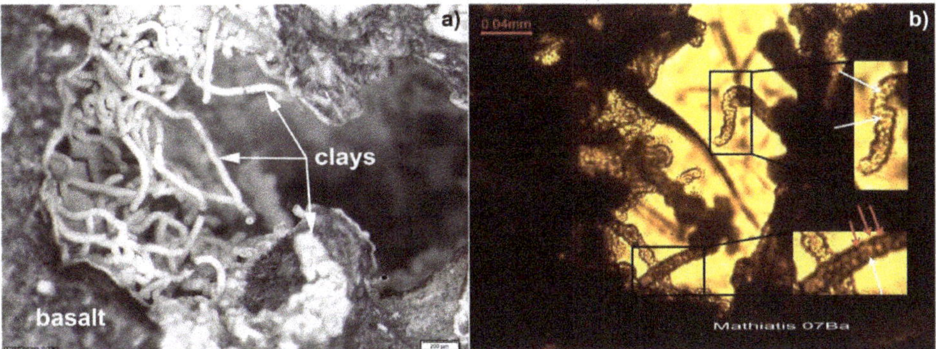

Figure 2. Microphotographs of filamentous structures in sample 07A in: (**a**) open vesicles, filamentous structures mineralized by white clay (white arrows) on basalt and; (**b**) calcite filled vein. The filaments protrude from the host rock/pore space interface, are curvi-linear and occur in abundance, forming complex networks. Filaments partitioned by septa-like walls marked by red arrows in 2b (insets). The septation is mostly repetitive and divide the filaments in individual compartments, and in a few filaments a nucleus occur in the center of each compartment marked by white arrows in 2b insets.

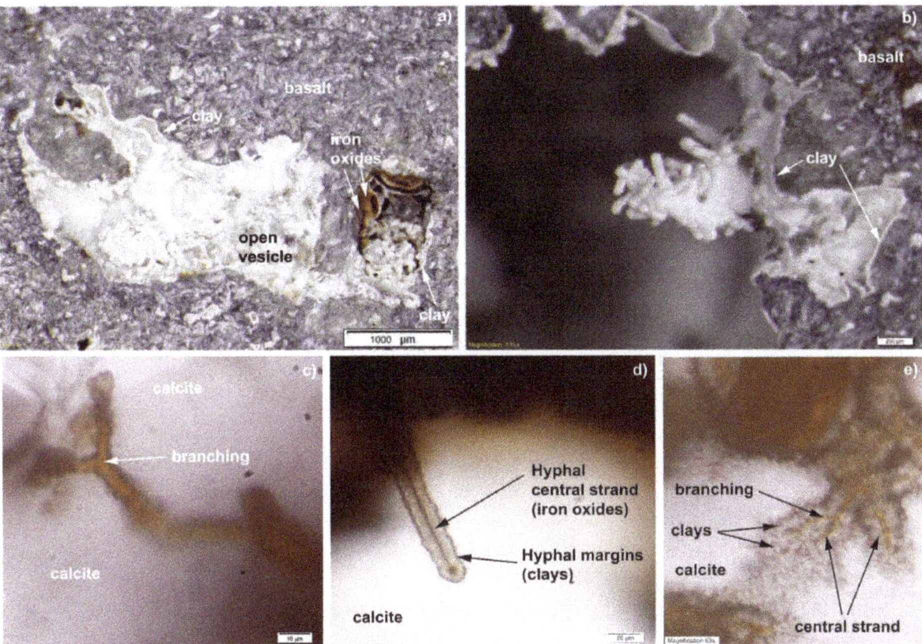

Figure 3. Microphotographs from sample 07A showing: (**a**) and (**b**) filaments originating from a clay/iron oxide film lining the interior of the pore space; (**c**) branching of a filament; (**d**) hyphal central strand and hyphal margins; (**e**) branching of the central strand following a general branching of the filaments and mineralizations of clays surrounding the filaments.

Figure 4. Microphotographs from sample 07A of filaments in open pore space with a terminal precipitation of zeolites forming a swelling at the filament tip.

Raman analysis of the film lining the wall interiors, the filament margins and interiors showed discernible peaks at 190, 265, 420 (with one shoulder at 360), 550, 615 and 700 cm^{-1} together with the strong wide band around 3600 cm^{-1} (Figure 5). The wavenumbers are close to published data for typical montmorillonite [(Na, Ca)$_{0.33}$(Al, Mg)$_2$(Si$_4$O$_{10}$)(OH)$_2$*nH$_2$O] with strong bands near 200, 425 (broad band) and 700 cm^{-1} with a less intense peak around 270 cm^{-1} together with a wide band around 3600 cm^{-1} [24,25]. The peaks at 190, 265, 420 and 700 cm^{-1} arise from SiO$_4$ tetrahedral unit vibrations. The peak at 550 cm^{-1} and the downshifted 270 cm^{-1} peak to 265 cm^{-1} indicates an Fe-rich composition, the peak at 420 cm^{-1} is influenced by an Al-rich composition and the peak that occur at the shoulder at 360 cm^{-1} is an indicator for some Mg [25]. The extra band at 615 cm^{-1} is usually not found in montmorillonite spectra, but matches the 604 cm^{-1} peak caused by interference of Fe^{3+} with Si-O-Si groups in the spectrum of glauconite [(K, Na)(Fe^{3+}, Al, Mg)$_3$(Si, Al)$_4$O$_{10}$(OH)$_2$] [25]. This extra band may suggest that the Fe-rich nature of the montmorillonite is due to the presence of a glauconite-like structure formed by replacement reactions of the montmorillonite. A mixture of smectite and mica is a common clay association in marine sediments [26] and may also explain why the measured Raman spectra are identified as montmorillonite and not nontronite. Good Raman spectra of clays may be problematic to obtain owing to the ultra-fine texture of the material and low degree of crystallinity [25]. In places, small fragments of rutile appear in the clay (Figure 5). The wide band around 3600 cm^{-1} is assigned to OH vibrations (Figure 5). The appearance of an additional strong peak at 3075 cm^{-1} in the filaments interiors compared to the margins suggest the presence of hydrocarbons and bands in the range 3000–3100 cm^{-1} that can be assigned to C-H stretches in aromatic compounds [27] or =CH$_2$ [28] (Figure 5). The central strand is mineralized by goethite with distinct Raman peaks at 246, 300, 386, 483 and 550 cm^{-1} in the interior of the structures (Figure 6). The presence of hydrocarbons and goethite in the interiors are displayed by a colour variation in optical microscopy. Filaments with compartments partitioned by repetitive septum-like walls and globular nuclei of 1 or a few µm show a content with a few wt % P according to EDS (Figure 7).

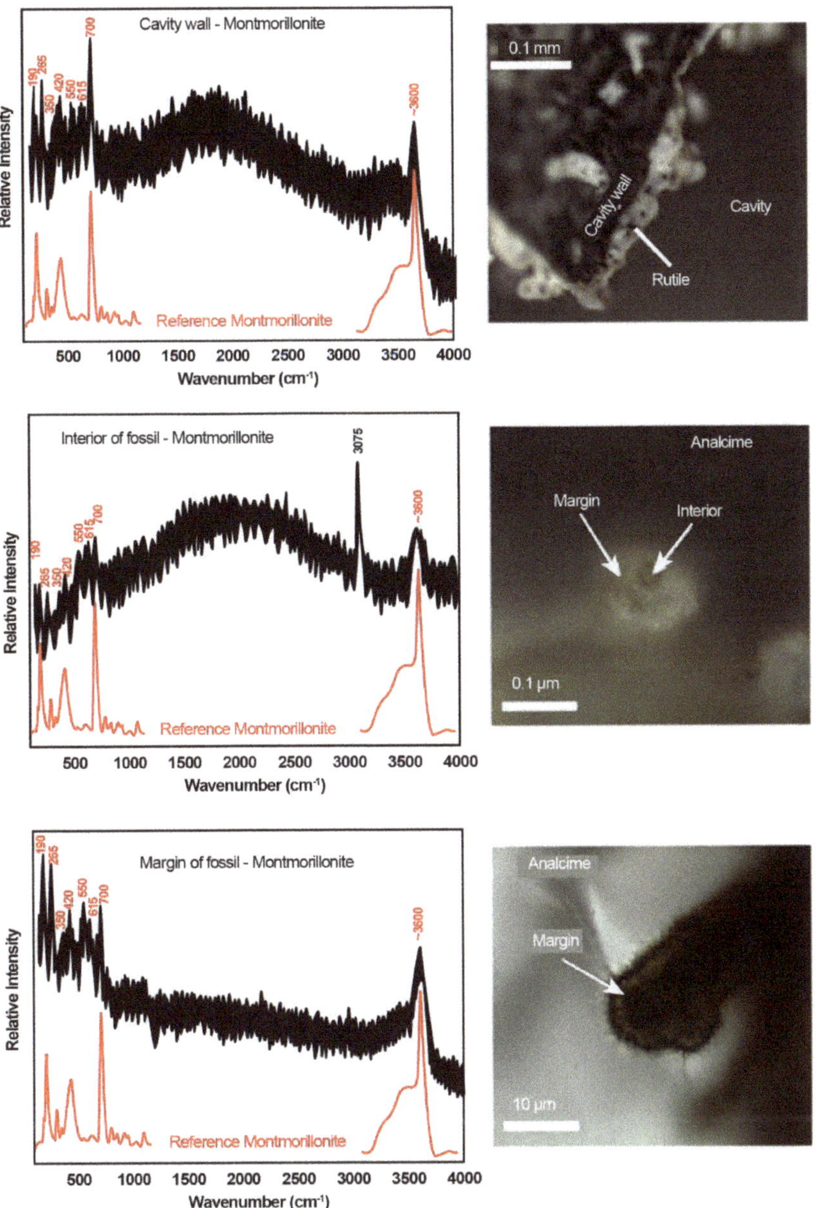

Figure 5. Raman spectra in the spectral range 100–4000 cm^{-1} of a cavity wall and filament structure in an analcime-filled vesicle in sample 07A. The spectra with peak positions in red numbers show that both the cavity wall and the filaments are composed of a Fe-rich montmorillonite. The spectra from the interior of the filament indicate that hydrocarbons are present with the peak at 3075 cm^{-1}. Reference spectra of montmorillonite after [25] have been incorporated in each spectral diagram. Rutile grains are incorporated in the cavity wall montmorillonite.

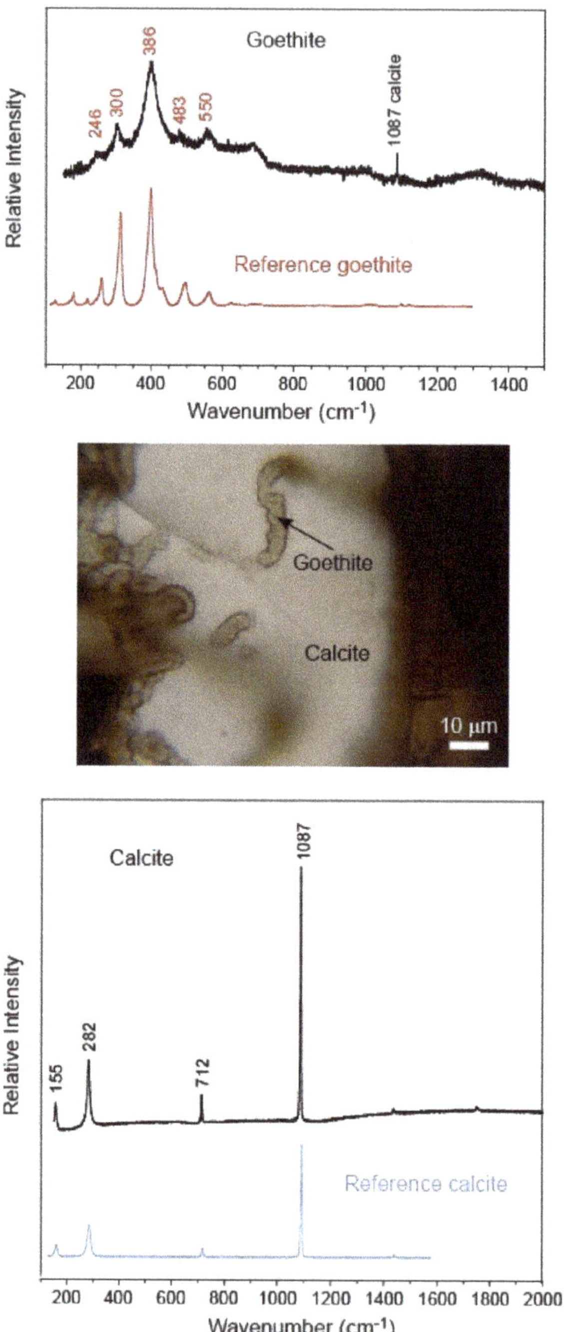

Figure 6. Upper diagram (sample 07A): Raman spectrum in the spectral range 100–1500 cm^{-1} of the interior of the filament in a calcite-filled vesicle. The spectrum shows peaks that are attributed to goethite. Lower diagram (sample 07A): Raman spectrum in the spectral range 100–2000 cm^{-1} of a calcite-filled vesicle. Reference spectra of goethite and calcite after Downs (2006) [22] have been incorporated in the figure.

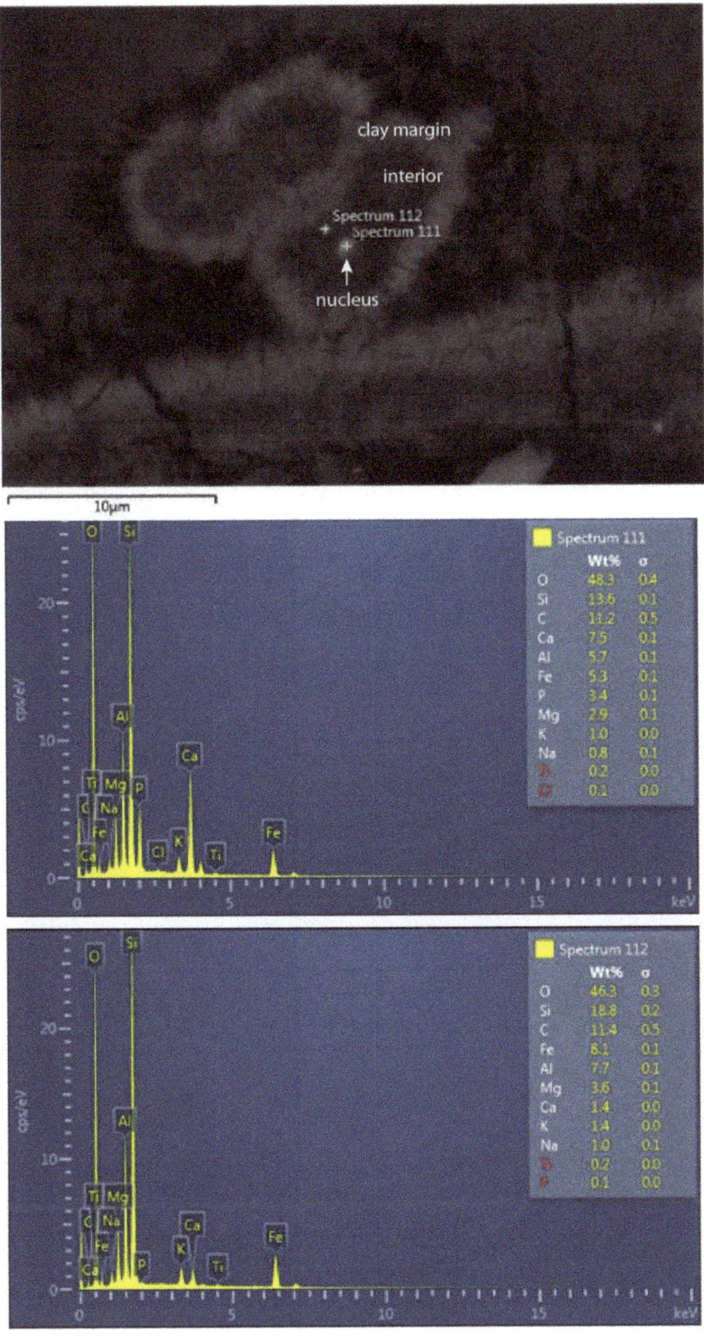

Figure 7. ESEM image from sample 07A with accompanying EDS spectra of filaments in cross section showing a phosphorus-rich nucleus surrounded by the overall clay in the filament interior.

Both the film linings and the filaments are seen in association with partially altered orthoclase laths in the vesicles (Figure 8a). The acquired Raman spectra of altered orthoclase (Figure 8b) reflect a gradual transition from orthoclase [$KAlSi_3O_8$] to analcime [$NaAlSi_2O_6*H_2O$]. The strongest Raman bands from the orthoclase that appear at 513 cm^{-1}, 475 cm^{-1} with a shoulder at 455 cm^{-1} and 285 cm^{-1} are believed to result from the main T-O-T and O-T-O vibration modes where T = Si and/or Al [29]. Other weaker bands (~755, ~805 and 1124 cm^{-1}) are also associated with structural changes in the T-O-T and O-T-O region due to varying bond lengths and angles in the tetrahedral crystal structure [29]. A weaker band at 157 cm^{-1} is assigned to Si-O vibrations involving larger cations like K and Na (Me-O in Figure 8b) in the structure [30]. In spectra A–E (Figure 8b) it can be seen that the replacement by analcime causes the band at 513 cm^{-1} to disappear and the band at 475 cm^{-1} to shift to a higher wavenumber at 482 cm^{-1}. The band at 455 cm^{-1} shifts towards a lower wavenumber at 389 cm^{-1} in analcime, whereas the moderately strong band at 285 cm^{-1} for orthoclase shifts upwards to 299 cm^{-1}. In analcime, an additional band at 3560 cm^{-1} appear that is assigned to O-H stretching of structurally bonded water [31]. The intensity of this band is gradually increasing from the intermediate phase in spectrum B and absent in the unaltered part (A) of the orthoclase. Areas in the altered orthoclase that have suffered different degrees of transition towards analcime are illustrated in Figure 8b where areas A-B match the corresponding spectra A–E. One larger calcite crystal shows distinct twinning, indicating temperatures <200 °C [32].

3.2. Fluid Inclusions

Fluid inclusions are in general rare in the samples but six inclusions in sample 07B with an aqueous liquid and a vapor bubble were distinguished in one specific sample of calcite (Figure 9). Five of these that were found in calcite veinlets, varied in size from 5 to 20 µm and were irregular with angular shapes. Such inclusions may have suffered post-entrapment modifications like stretching or leakage, which can result in a shift to higher homogenization temperatures. The recorded homogenization temperatures from these inclusions, 109 °C to 131 °C (to liquid), should be used with caution and are probably somewhat too high. However, one large fluid inclusion (40 µm) that was found in a cavity filled with well-preserved calcite has survived without leakage. The small size of the cavity and the surrounding basaltic host rock has protected the fluid inclusion from deformation and leakage. Homogenization temperature of this inclusion was measured at 75 °C (to liquid) which is a minimum value for the formation temperature. Initial melting of all six inclusions was observed at temperatures around −35 °C indicating a mixed composition dominated by Mg^{2+}, Fe^{2+}, Ca^{2+} and Na^+ [23]. Final ice melting occurred between −2.4 °C and −2.7 °C, corresponding to salinities from 4.0 to 4.4 eq. mass % NaCl [33].

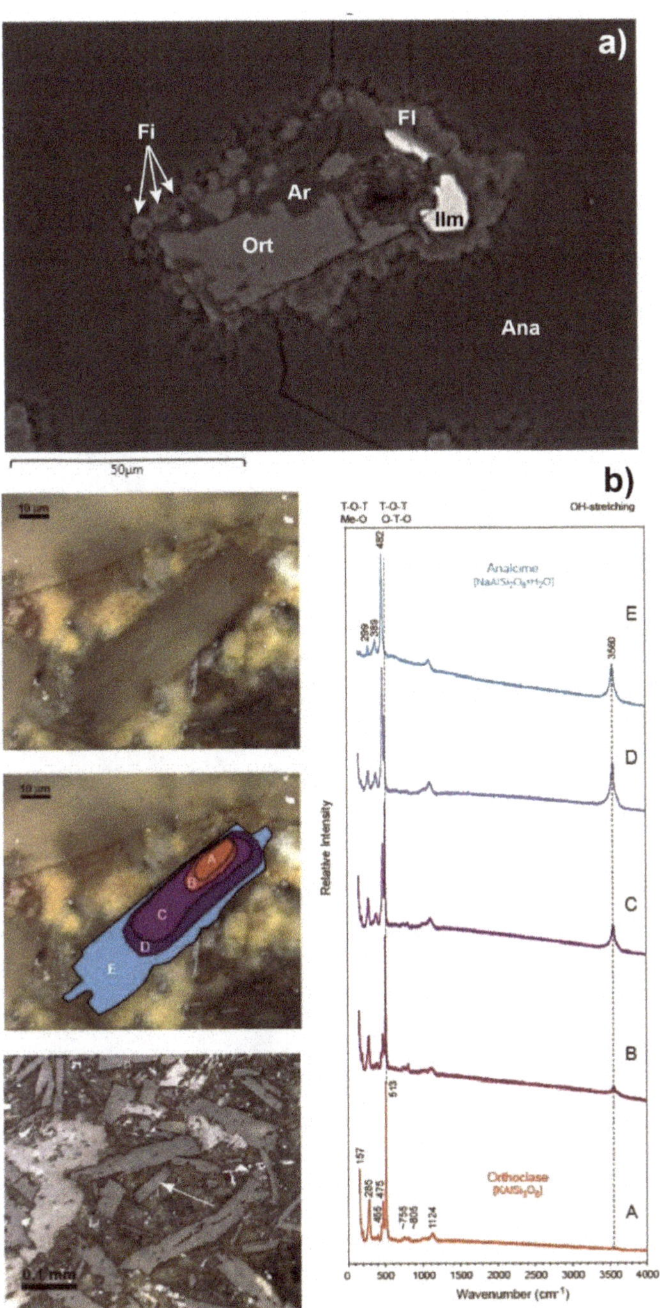

Figure 8. (**a**) ESEM image from sample 07A of a orthoclase (Ort) and ilmenite (Ilm) crystal with alteration rims (Ar) and associated filaments (Fi) as well as film lining (Fl) the mineral surfaces embedded in an analcime (Ana); (**b**) Raman spectra in the spectral range 100–4000 cm^{-1} of an altered orthoclase crystal shown in the uppermost microphotograph and its position in the sample (lowermost microphotograph). The Raman spectra A–E illustrate the transition from the endmember feldspar identified as orthoclase (red spectrum A) to the endmember zeolite identified as analcime (blue spectrum E) with intermediate

phases represented by red- to blueviolet spectra B to D. Band positions are indicated by wavenumbers (cm^{-1}) on the red spectrum for orthoclase and on the blue spectrum for analcime. The spectral range for vibrational bands Me-O (Me = K, Na), T-O-T and O-T-O (T = Si, Al), and OH is given on top of the diagram. The spectra in the diagram correspond to the areas marked A–E (with the same color) in the middle microphotograph. These areas represent different zones of the gradual transition from the original orthoclase still left in the core to the final phase analcime along the margin of the crystal.

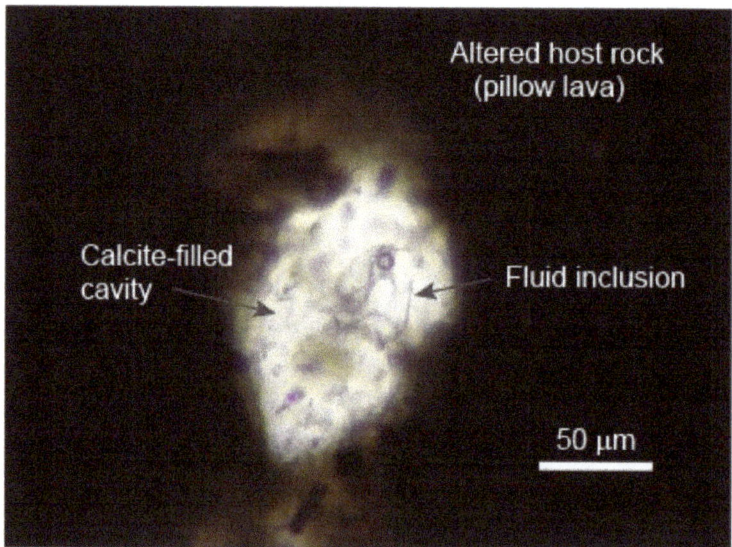

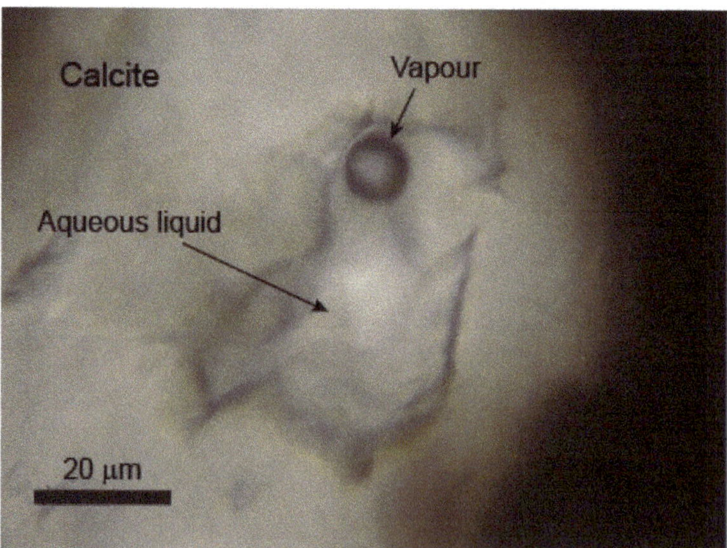

Figure 9. Upper panel: Microphotograph from sample 07B of a large aqueous fluid inclusion in a calcite-filled cavity. Lower panel: close up of the same fluid inclusion.

4. Discussion

4.1. Biogenicity

The following criteria are used to test the biogenicity of the possible biological textures in the pillow lavas: 1) the geological context, 2) the syngenicity and indigenousness to the rock and secondary minerals, 3) the morphology, and 4) the composition of the putative fossils [21,34–36].

1) The filamentous structures are found in both open and filled veins and vesicles in pillow lavas that represent ocean crust of Cretaceous age [19]. Similar filamentous structures has been described from both modern oceanic crust [7,10,12,37] and ophiolites [38,39], and interpreted as fossilized microorganisms. In previous papers, volcanic glass in pillow lavas from Troodos have been shown to host ichnofossils interpreted as morphological remains of microbial activity that occurred prior to the oceanic crust was pushed onto the African plate [17,18]. Furthermore, the oceanic crust is known to host a substantial deep biosphere [12,40], thus, the geological context of the samples is compatible with microbial life.

2) The filaments protrude from the montmorillonite film lining the walls of the pore space and both consist of the same clay, thus the film and the filaments are contemporaneous. No secondary mineralization occurs between the vesicle/vein walls and the suggested microfossils, thus the putative microorganisms colonized the rock during the first alteration stage prior to the formation of the vein-filling carbonates. The filaments are totally mineralized by clays and iron oxides formed by circulating hydrothermal fluids (see Sections 4.2–4.4), and are thus not modern contaminants. Where the filaments are encapsulated by carbonates or zeolites, the secondary mineralization post-dates the growth of the putative microorganisms. Precipitation of secondary minerals and subsequent filling of the voids normally happens 10–20 Ma after formation of the host rock [41]. The ophiolite was emplaced onto land some 20 Ma ago, giving a minimum age of the putative fossils [20] indicating that the microorganisms are syngenetic to early hydrothermal fluids rather than being a modern colonization.

3) No abiotic process is known to form micro-structures with morphologies and growth patterns as those described here. However, there is a substantial literature on microbial fossils in oceanic crust and ophiolites identical to the current structures [12,38] and the references therein. The initial forming of a film lining the interior of the vein/vesicle walls, the subsequent perpendicular growth from the walls into the open pore space, and mineralization by clays and iron oxides is identical to previously described fossilized microorganisms from subseafloor crust and ophiolites, of which most have been interpreted as remains of filamentous endolithic fungi [7,12,37–39,42]. Also, the curvi-linear appearance, the branching, occurrence of a central strand and repeated septations are typical for such filamentous microfossils. No conidia was observed. The width of the filaments ranging between 5 and 50 μm cannot be used as a reliable criterion for biogenicity or to discriminate between prokaryotes or eukaryotes since big bacteria are known to have diameters up to 1 mm [43]. Fungal hyphae are known to have diameters between 2–27 μm, but fossilized hyphae from the ocean floor have been shown to possess wider diameters at times [8,10]. No obvious mineral growth during diagenesis can be seen on the filaments, but due to the thick diameters, it cannot be entirely ruled out. Many of the filaments contains a central strand mineralized by iron oxides and organic matter according to Raman spectroscopy. This could either represent shrinkage of the cytoplasm during fossilization or the remains of thick cell walls [37]. Occasionally, the filaments have repetitive septa resulting in cell-like compartmentalization. Spherical micron-sized textures are found in these compartments, usually containing P (detailed description under point 4). This could also either represent shrinkage of the cytoplasm during fossilization or the possible remains of a cell nucleus. Due to the defined spherical shape, the coherence throughout each compartment, and the exclusive presence of P to the nucleus we suggest the latter. The presence of a viable cell nucleus would exclude a prokaryotic interpretation and be in favor of a eukaryotic interpretation. The filaments appear to have a mycelium-like network, similar to that found among fungi [44], and more or less identical to previously described fossilized fungal mycelium in the igneous oceanic crust [8,10,12]. Actinobacteria are the only known prokaryotes

that form mycelia but have diameters up to 2 μm, thus, are excluded due to the large diameters among the current filaments [45]. In the absence of a known abiotic explanation, and the consistency with a well-known fossil record of the oceanic crust [12], and the references therein, a biological explanation for the morphological characteristics of the current microstructures are the most conceivable.

4) EDS and Raman show that the filaments consist of smectite with a composition of Fe-rich montmorillonite corresponding to fossilization of microorganisms in basaltic crust [12]. EDS also shows P enrichment in possible cell-like interiors, which could be biologically derived. Phosphorous is one of few elements required for all life regarding energy acquirement, genetic information and building membranes [46–48]. Cells containing P are usually consumed by microorganisms during diagenesis. Oxic waters, however, promote preservation of P both through the formation of insoluble P compounds, but also through adsorption and co-precipitation of P onto ferrihydrites [49]. The latter could explain why P has been preserved in the iron-rich central strands.

Some filaments have a carbonaceous content preserved in the central core, according to the Raman analysis (Figure 3). This carbonaceous material is believed to be indigenous to the filaments, and thus represent remains of primary organic matter, which would suggest a biological origin.

The above criteria for biogenicity have been addressed and successfully fulfilled in favor of a biological interpretation. It is further suggested, based on the morphological features that the filaments represent fossilized fungal hyphae rather than filamentous prokaryotes.

4.2. Hydrothermal Conditions of the Habitats

Modern hydrothermal vent areas have temperatures between 350 °C–400 °C along the active axial ridge, which decreases gradually from the spreading center [50]. For Troodos, discharge temperatures were between 300 °C–350 °C [51]. One larger calcite from the Mathiatis mine shows distinct twinning, indicating temperatures <200 °C [32]. Homogenization temperatures of fluid inclusions in the calcite from the Mathiatis mine further show temperatures >75 °C. Calcite formation in the sheeted dikes on Cyprus has been estimated to 75 °C–100 °C and is considered to be somewhat higher than calcite formation in the lavas [52]. Zeolite and smectites indicate temperatures <100 °C, and mordenite also indicates temperatures <100 °C [53]. Celadonite is usually formed at temperatures <50 °C, where a previous study has given a formation temperature between 15 °C–20 °C for celadonite on Cyprus, and celadonite found together with saponite in basaltic rocks showed formation temperatures between 50 °C–90 °C [54]. The Raman analysis showed poorly crystalline carbon in the fossilized fungi, which indicates temperatures <150 °C [55–58].

Hydrothermal alteration of the Troodos pillow lavas is given by celadonite and Fe-rich oxides, followed by saponite seen in the vesicles and veins. This stage of alteration should have occurred at temperatures <50 °C for celadonite, and <75 °C for saponite. This sequence is seen in most of the samples, though celadonite is sparse, and saponite is much more abundant. Secondary filling suggests two additional later events with hydrothermal fluids, where the first precipitated Na or Ca zeolites with temperatures <100 °C. The last event precipitated Ca carbonates with temperatures >75 °C.

The organisms thrived in open vesicles prior to secondary precipitation of zeolite or calcite. Thus, this gives a temperature window ranging from ~50 to 75 °C in which the microorganisms could have occurred. The lower range of these temperatures are within the upper temperature maximum for fungi, which is 62 °C [59], while the higher temperature range is above the fungal temperature maximum. This suggests that temperatures partially were within the known temperature limits for fungi during colonization. Furthermore, calcite forms as fluid temperatures increase in the system, and previous studies combining microfossils and fluid inclusions have shown that the temperatures were lower, and within the temperature window for life, in between hotter, mineral-forming pulses of fluids [60].

Basaltic fluids have a general pH between 8–10, but hydrothermal alterations can change the pH in the host rock, where decreasing Mg increases the pH [61,62] Cold oceanic water has a pH between 7–8, which mixes with the magmatic fluids that have a pH between 4–6 when it percolates down into the oceanic floor [63]. During its rise to the ocean floor, pH and/or redox changes will precipitate the elements in dissolution, creating new minerals. The high clay content in all the samples indicates pH higher than 5 [64]. Mordenite is usually found in environments with a pH between 7–9 [65], analcime is found in environments with pH higher than 8 [66], while Ca carbonates are found in environments with pH higher than 8–9 [67]. Living fungi in aerobic environments have been cultivated and show a preference for pH around 4.5–8.3 [68], and studies of living fungi from subterranean environments shows that they can survive in extreme environments with pH up to 10 [69].

4.3. Mobilization of Elements

K and Fe are two of the most abundant cations found in all living organisms. They help control biochemical functions and growth, and the uptake of these ions is thus essential for the microorganisms to survive [70,71]. In the samples, fossilized microorganisms are often found in direct contact with interstitial K-feldspar and Fe oxides (Figure 8b). The Fe-oxides that are in direct contact with the fossils shows no recrystallization along the edges, which indicates either none or Fe-dissolution on a smaller scale than our methods could discern. Recrystallization of K-feldspar at the edges is due to mobilization of K^+ and recrystallization with Na^+, either abiotically or biologically mediated. The close association with the fossils suggest the latter. Fungi are known to secrete chelating compounds such as siderophores or organic acids to dissolve minerals [72,73]. Biofilm formation is a microbial strategy to initiate microbe–mineral interactions and control micro-environments including surface properties and surface charges [74,75].

The high abundance of fossils and the fact that the samples are from a mine may be due to higher metal content, as well as higher temperatures of the hydrothermal fluids, giving rise to more dissolved ions from the host rock in general. However, the high abundance of putative fungal fossils in the ores, and the connection to a high metal content is probably indirect. Chemoautotrophic prokaryotes favor metal availability and would therefore fix carbon from the CO_2 in the fluids, while oxidizing metals that are available in the hydrothermal system. A higher abundance of prokaryotic biomass would mean an increased pool of carbohydrates available for the heterotrophic fungi. Thus, the high fungal abundance in the ore might be a secondary result of a higher chemoautotrophic activity closer to the hydrothermal system. The higher abundance of fungi could also be due to better preservation since more elements for fossilization is available with increasing hydrothermal activity. We believe that with increasing metal content and higher temperature of the hydrothermal fluids (i.e., more dissolved elements), a combination of both these possibilities is likely.

4.4. Fossilization

Fossilization of the microorganisms by clays and iron oxides occurred as elements in the fluids successively replaced the organic material. Subsequent mineralization precipitated within or between the cells in the organisms, leaving remains of organic matter in the fossil interiors as revealed by the Raman analysis. There is a known connection between clay minerals and the polymerization of biomolecules, as well as complex biopolymers [72,76–78]. Biomolecules can adhere to vacant sites in clay minerals and there is a molecular structure between the both that favors this coupling. Ivarsson et al. (2013) [7] argued that the fossilization process of subseafloor fungi start while the organisms were still alive, based on the non-dehydrated morphology lacking among dead fungi [79]. During fossilization of the microorganisms in the current study, as well as in previous studies [7,37], this coupling may be of importance in describing how organic molecules can recrystallize into clays. Hydrocarbons contain single bonds that are easy to break. The element-rich fluids that enter the host rock should be able to start recrystallizing the organic molecules by breaking the single bond between the carbon and hydrogen, opening a negative carbon site. Montmorillonite smectite was found as the more common fossilization mineral of the fungi. This is a 2:1 structured clay mineral that consists of 2 repetitive TOT layers. Because of the layered structure in clays, elements are attracted by weak electrostatic bonds making it easy to build the next layer. The TOT structure has an overall negative charge, which is balanced by inter-layered cations between the TOT layers, usually by cations like K, Na, Ca and water molecules.

EDS and Raman analysis of hyphae in cross-section show that the clay ranges from Mg-rich in the core to Fe- and K-rich in the outer wall, with enrichment of Ti in between. Since K and Fe is thought to have been absorbed by the living organisms, these cations are readily available in situ, while Mg, Ti, Na and Ca most likely have been added by later hydrothermal fluids. All the elements, except Ti, are probably enriched from the seawater, or by the dissolution of the basalt. Mg, however, could also have been dissolved during the breakdown of the clinopyroxene and the olivine in the original host rock. Ti could be probably delivered with the magmatic water and precipitated when pH and temperature changes. Besides, clay and rutile crystals are found in vacant sites of the montmorillonite, and goethite is found in the central strand. The fossils appear to be localized around Fe-oxides, where Fe enrichment is also seen in the clays. Goethite in the fungi might therefore have been due to a high Fe content, and fossilization into an oxide was preferred prior to clay formation, or the fungi might have already deposited the oxide while it was alive. Similar oxide-rich strands have been found in other studies [10,12]. Based on the hydrothermal history, fossilization in these samples might have occurred in the following way:

Microorganisms are introduced into the system with the early low temperature hydrothermal fluids, responsible for the celadonite and saponite alteration. This is followed by a second hydrothermal event with Na, Ca and Si-rich fluids precipitating zeolites, increasing the general pH in the system, as well as the temperatures, and thus stressing the microorganisms to start precipitate clay and oxide minerals as a defense mechanism [7]. The last stage of hydrothermal fluids precipitates Ca carbonates, killing the microorganisms due to too high temperature and pH. Total fossilization probably occurred before total encapsulation, based on fossilized, partly or non-encapsulated microorganisms found in the samples (Figure 2e,f). Hyphae have zeolite precipitated at the tip, which could indicate that the charged surface of these microorganisms might have been a favorable nucleation site. Because the hyphae in the sample show the inner strand in the zeolite encapsulation; this indicates that these microorganisms were encapsulated first after fossilization occurred, since the filament is broken off showing the inner strand. Thus, fossilization seems to be highly dependent on both fluid composition as well as rate and time of secondary precipitation.

5. Conclusions

This study focuses on filamentous fossils from the Troodos ophiolite (91 Ma), Cyprus. Based on fungal characteristics such as size, growth behavior, hyphae, frequent branching, central strand, repetitive septa, and potential cell nucleus, a fungal interpretation was inferred for the fossils. Microbial colonization is found on K- and Fe-rich minerals, which they probably adhere to for elemental dissolution. Fossilization of the microorganisms has mainly been done by Fe- and Mg-rich clays, where some fungi have been fossilized by goethite in the central strand, and rutile crystals in vacant sites of the montmorillonite. Elements such as Mg, Ca, and Na have been introduced with the oceanic water, and Ti with the magmatic water. Fossilization was initiated during temperature and pH changes of later hydrothermal activity, where at least three hydrothermal events can be identified. An early hydrothermal event was initiated with temperatures <75 °C, precipitating celadonite and saponite in open vesicles and veins, as well as introduced microbial life into open pore spaces. A second event with temperatures <100 °C precipitated Na and Ca zeolites, increasing the pH to 7–9, stressing the microorganisms into starting to adhere clays as protection. Fossilization was finalized with a last hydrothermal event with temperatures >75 °C, precipitating Ca carbonates, increasing pH to >8–9, as well as making the environment inhabitable for the microorganisms. We, thus, suggest the hydrothermal fluids, temperature, as well as the rate and time of secondary precipitation in open vesicles and veins are the main factors that control preservation.

Author Contributions: Conceptualization, D.-T.C., M.I. and A.N.; formal analysis, D.-T.C.; investigation, D.-T.C.; writing—original draft preparation, D.-T.C.; writing—review and editing, M.I. and A.N.; project administration, M.I. and A.N.

Funding: This research was funded by the Swedish research council, grant number 2017-04129 and 2013-7320, the Swedish National Space Agency, grant number 100/13, and a Villum Investigator Grant to Don Canfield, grant number 16518.

Acknowledgments: The authors acknowledge Marianne Ahlbom, Stockholm University for assistance with the EDS and SEM analyses, and Curt Broman, Stockholm University, for Raman spectroscopy and fluid inclusion analyses.

Conflicts of Interest: The authors declare no conflict of interest.

References

1. Schrenk, M.O.; Huber, J.A.; Edwards, K.J. Microbial Provinces in the Subseafloor. *Ann. Rev. Mar. Sci.* **2009**, *2*, 279–304. [CrossRef] [PubMed]

2. Orcutt, B.N.; Bach, W.; Becker, K.; Fisher, A.T.; Hentscher, M.; Toner, B.M.; Wheat, C.G.; Edwards, K.J. Colonization of subsurface microbial observatories deployed in young ocean crust. *ISME J.* **2011**, *5*, 692. [CrossRef] [PubMed]

3. Mason, O.U.; Nakagawa, T.; Rosner, M.; Van Nostrand, J.D.; Zhou, J.; Maruyama, A.; Fisk, M.R.; Giovannoni, S.J. First Investigation of the Microbiology of the Deepest Layer of Ocean Crust. *PLOS ONE* **2010**, *5*, e15399. [CrossRef] [PubMed]

4. Orcutt, B.; Wheat, C.G.; Edwards, K.J. Subseafloor Ocean Crust Microbial Observatories: Development of FLOCS (FLow-through Osmo Colonization System) and Evaluation of Borehole Construction Materials. *Geomicrobiol. J.* **2010**, *27*, 143–157. [CrossRef]

5. Lever, M.A.; Rouxel, O.; Alt, J.C.; Shimizu, N.; Ono, S.; Coggon, R.M.; Shanks, W.C.; Lapham, L.; Elvert, M.; Prieto-Mollar, X.; et al. Evidence for microbial carbon and sulfur cycling in deeply buried ridge flank basalt. *Science* **2013**, *339*, 1305–1308. [CrossRef]

6. Jungbluth, S.P.; Lin, H.-T.; Cowen, J.P.; Glazer, B.T.; Rappé, M.S. Phylogenetic diversity of microorganisms in subseafloor crustal fluids from Holes 1025C and 1026B along the Juan de Fuca Ridge flank. *Front. Microbiol.* **2014**, *5*, 119. [CrossRef]

7. Ivarsson, M.; Bengtson, S.; Skogby, H.; Belivanova, V.; Marone, F. Fungal colonies in open fractures of subseafloor basalt. *Geo-Marine Lett.* **2013**, *33*, 233–243. [CrossRef]

8. Ivarsson, M.; Peckmann, J.; Tehler, A.; Broman, C.; Bach, W.; Behrens, K.; Reitner, J.; Böttcher, M.E.; Ivarsson, L.N. Zygomycetes in Vesicular Basanites from Vesteris Seamount, Greenland Basin—A New Type of Cryptoendolithic Fungi. *PLOS ONE* **2015**, *10*, e0133368. [CrossRef]

9. Schumann, G.; Manz, W.; Reitner, J.; Lustrino, M. Ancient Fungal Life in North Pacific Eocene Oceanic Crust. *Geomicrobiol. J.* **2004**, *21*, 241–246. [CrossRef]

10. Bengtson, S.; Ivarsson, M.; Astolfo, A.; Belivanova, V.; Broman, C.; Marone, F.; Stampanoni, M. Deep-biosphere consortium of fungi and prokaryotes in Eocene subseafloor basalts. *Geobiology* **2014**, *12*, 489–496. [CrossRef]

11. Ivarsson, M.; Bengtson, S.; Neubeck, A. The igneous oceanic crust—Earth's largest fungal habitat? *Fungal Ecol.* **2016**, *20*, 249–255. [CrossRef]

12. Ivarsson, M.; Bengtson, S.; Skogby, H.; Lazor, P.; Broman, C.; Belivanova, V.; Marone, F. A Fungal-Prokaryotic Consortium at the Basalt-Zeolite Interface in Subseafloor Igneous Crust. *PLOS ONE* **2015**, *10*, e0140106. [CrossRef] [PubMed]

13. Bengtson, S.; Sallstedt, T.; Belivanova, V.; Whitehouse, M. Three-dimensional preservation of cellular and subcellular structures suggests 1.6 billion-year-old crown-group red algae. *PLoS Biol.* **2017**, *15*, e2000735. [CrossRef] [PubMed]

14. McMahon, S.; Parnell, J. The deep history of Earth's biomass. *J. Geol. Soc.* **2018**, *175*, 716–720. [CrossRef]

15. Ivarsson, M.; Holm, N.G.; Neubeck, A. The deep biosphere of the subseafloor igneous crust. In *Handbook of Environmental Chemistry*; Demina, L., Galkin, S., Eds.; Springer: Cham, Switzerland, 2016; Volume 50.

16. Osozawa, S.; Shinjo, R.; Lo, C.-H.; Jahn, B.-M.; Hoang, N.; Sasaki, M.; Ishikawa, K.; Kano, H.; Hoshi, H.; Xenophontos, C.; et al. Geochemistry and geochronology of the Troodos ophiolite: An SSZ ophiolite generated by subduction initiation and an extended episode of ridge subduction? *Lithosphere* **2012**, *4*, 497–510. [CrossRef]

17. Furnes, H.; Muehlenbachs, K.; Tumyr, O.; Torsvik, T.; Xenophontos, C. Biogenic alteration of volcanic glass from the Troodos ophiolite, Cyprus. *J. Geol. Soc.* **2001**, *158*, 75–84. [CrossRef]

18. Furnes, H.; Banerjee, N.R.; Staudigel, H.; Muehlenbachs, K.; McLoughlin, N.; De Wit, M.; Van Kranendonk, M. Comparing petrographic signatures of bioalteration in recent to Mesoarchean pillow lavas: Tracing subsurface life in oceanic igneous rocks. *Precambrian Res.* **2007**, *158*, 156–176. [CrossRef]

19. Pearce, J.A.; Lippard, S.J.; Roberts, S. Characteristics and tectonic significance of supra-subduction zone ophiolites. *Geol. Soc. London, Spec. Publ.* **2008**, *16*, 77–94. [CrossRef]

20. Moores, E.M.; Vine, F.J. The Troodos Massif, Cyprus and other Ophiolites as Oceanic Crust: Evaluation and Implications. *Philos. Trans. R. Soc. A: Math. Phys. Eng. Sci.* **1971**, *268*, 443–467. [CrossRef]

21. Ivarsson, M. Advantages of doubly polished thin sections for the study of microfossils in volcanic rock. *Geochem. Trans.* **2006**, *7*, 5. [CrossRef]

22. Downs, R.T. The RRUFF Project: An integrated study of the chemistry, crystallography, Raman and infrared spectroscopy of minerals. In *Program and Abstracts of the 19th General Meeting of the International Mineralogical Association in Kobe, Japan, 2006*; International Mineralogical Association: Kobe, Japan, 2006.

23. Shepherd, T.J.; Rankin, A.H.; Alderton, D.H. *A Practical Guide to Fluid Inclusion Studies*; Chapman & Hall: London, UK, 1985; p. 239.

24. Bishop, J.L.; Murad, E. Characterization of minerals and biogeochemical markers on Mars: A Raman and IR spectroscopic study of montmorillonite. *J. Raman Spectrosc.* **2004**, *35*, 480–486. [CrossRef]

25. Wang, A.; Freeman, J.J.; Jolliff, B.L. Understanding the Raman spectral features of phyllosilicates. *J. Raman Spectrosc.* **2015**, *46*, 829–845. [CrossRef]

26. Charpentier, D.; Buatier, M.; Jacquot, E.; Gaudin, A.; Wheat, C. Conditions and mechanism for the formation of iron-rich Montmorillonite in deep sea sediments (Costa Rica margin): Coupling high resolution mineralogical characterization and geochemical modeling. *Geochim. Cosmochim. Acta* **2011**, *75*, 1397–1410. [CrossRef]

27. Orange, D.; Knitile, E.; Farber, D.; Williams, Q. Raman spectroscopy of crude oils and hydrocarbon fluid inclusions: A feasibility study. *Miner. Spectrosc.* **1996**, *5*, 65–81.

28. Lin-Vien, D.; Colthup, N.B.; Fateley, W.G.; Grasselli, J.G.; Lin-Vien, D.; Colthup, N.B.; Fateley, W.G.; Grasselli, J.G. *The Handbook of Infrared and Raman Characteristic Frequencies of Organic Molecules*; Elsevier Science: Amsterdam, The Netherlands, 1991; ISBN 0080571166, 9780080571164.

29. Freeman, J.J.; Wang, A.; Kuebler, K.E.; Jolliff, B.L.; Haskin, L.A. Characterization of Natural Feldspars by Raman Spectroscopy for Future Planetary Exploration. *Can. Miner.* **2008**, *46*, 1477–1500. [CrossRef]

30. Frogner, P.; Broman, C.; Lindblom, S. Weathering detected by Raman spectroscopy using AI-ordering in albite. *Chem. Geol.* **1998**, *151*, 161–168. [CrossRef]

31. Frost, R.L.; Lopez, A.; Theiss, F.L.; Romano, A.W.; Scholz, R. A vibrational spectroscopic study of the silicate mineral analcime—Na2(Al4SiO4O12)·2H2O—A natural zeolite. *Spectrochim. Acta Part A: Mol. Biomol. Spectrosc.* **2014**, *133*, 521–525. [CrossRef]

32. Burkhard, M. Calcite twins, their geometry, appearance and significance as stress-strain markers and indicators of tectonic regime: a review. *J. Struct. Geol.* **1993**, *15*, 351–368. [CrossRef]

33. Bodnar, R.J. *Fluid Inclusions—Analysis and Interpretation*; Anderson, A., Marshall, D., Eds.; Mineralogical Association of Canada: Quebec City, QC, Canada, 2003.

34. Gibson, E.; McKay, D.; Thomas-Keprta, K.; Wentworth, S.; Westall, F.; Steele, A.; Romanek, C.; Bell, M.; Toporski, J. Life on Mars: evaluation of the evidence within Martian meteorites ALH84001, Nakhla, and Shergotty. *Precambrian Res.* **2001**, *106*, 15–34. [CrossRef]

35. Buick, R. Microfossil Recognition in Archean Rocks: An Appraisal of Spheroids and Filaments from a 3500 M.Y. Old Chert-Barite Unit at North Pole, Western Australia. *Palaios* **1991**, *5*, 441–459. [CrossRef]

36. Schopf, J.W.; Walter, M.R. Archean microfossils: New evidence of ancient microorganisms. In *Earth's Earliest Biosphere, Its Origin and Evolution*; Schopf, J.W., Ed.; Princeton University Press: Princeton, NJ, USA, 1983.

37. Ivarsson, M.; Bengtson, S.; Belivanova, V.; Stampanoni, M.; Marone, F.; Tehler, A. Fossilized fungi in subseafl oor Eocene basalts. *Geology* **2012**, *40*, 163–166. [CrossRef]

38. Peckmann, J.; Bach, W.; Behrens, K.; Reitner, J. Putative cryptoendolithic life in Devonian pillow basalt, Rheinisches Schiefergebirge, Germany. *Geobiology* **2008**, *6*, 125–135. [CrossRef] [PubMed]

39. Eickmann, B.; Bach, W.; Kiel, S.; Reitner, J.; Peckmann, J. Evidence for cryptoendolithic life in Devonian pillow basalts of Variscan orogens, Germany. *Palaeogeogr. Palaeoclim. Palaeoecol.* **2009**, *283*, 120–125. [CrossRef]

40. Orcutt, B.N.; Sylvan, J.B.; Knab, N.J.; Edwards, K.J. Microbial Ecology of the Dark Ocean above, at, and below the Seafloor. *Microbiol. Mol. Boil. Rev.* **2011**, *75*, 361–422. [CrossRef]

41. Pirajno, F. Hydrothermal Processes and Mineral Systems. *Hydrothermal Processes and Mineral Systems* 2009.

42. Ivarsson, M.; Bach, W.; Broman, C.; Neubeck, A.; Bengtson, S. Fossilized Life in Subseafloor Ultramafic Rocks. *Geomicrobiol. J.* **2018**, *35*, 460–467. [CrossRef]

43. Schulz, H.N.; Jørgensen, B.B. Big Bacteria. *Ann. Rev. Microbiol.* **2002**, *55*, 105–137. [CrossRef]

44. Webster, J.; Weber, R. *Introduction to Fungi*, 3rd ed.; Cambridge University Press: Cambridge, UK, 1981; ISBN 9780521807395.

45. Goodfellow, M.; Kämpfer, P.; Busse, H.J.; Trujillo, M.E.; Suzuki, K.I.; Ludwig, W.; Whitman, W.B. (Eds.) *Busse Bergey's Manual of Systematic Bacteriology*; Springer: New York, NJ, USA, 2012; Volume 5.

46. Crosby, C.H.; Bailey, J.V. The role of microbes in the formation of modern and ancient phosphatic mineral deposits. *Front. Microbiol.* **2012**, *3*, 241. [CrossRef]

47. Sindhu, S.S.; Phour, M.; Choudhary, S.R.; Chaudhary, D. Phosphorus Cycling: Prospects of Using Rhizosphere Microorganisms for Improving Phosphorus Nutrition of Plants. In *Geomicrobiology and Biogeochemistry*; Springer: Berlin, Germany, 2014; pp. 199–237.

48. Raven, P.; Johnson, G.; Mason, K.; Losos, J.; Singer, S. *Biology*, 11th ed.; McGraw-Hill: New York, NY, USA, 2017; ISBN 1259188132, 9781259188138.

49. Fru, E.C.; Hemmingsson, C.; Holm, M.; Chiu, B.; Iñiguez, E. Arsenic-induced phosphate limitation under experimental Early Proterozoic oceanic conditions. *Earth Planet. Sci. Lett.* **2016**, *434*, 52–63. [CrossRef]

50. Candela, P.A. Ores in the Earth's Crust. In *The Crust*; Rudick, R.L., Ed.; Elsevier Science: Amsterdam, The Netherlands, 2005; p. 702.

51. Spooner, E.T.C. Cu-pyrite mineralization and seawater convection in oceanic crust-the ophiolitic ore deposits of Cyprus. *Geol. Assoc. Can.* **1980**, *20*, 685–704.

52. Cann, J.; Gillis, K. Hydrothermal insights from the Troodos Ophiolite, Cyprus. In *Hydrogeology of the Oceanic Lithosphere*; Cambridge University Press: New York, NY, USA, 2004; pp. 272–310.

53. Pálmason, G.; Arnórsson, S.; Fridleifsson, I.B.; Kristmannsdóttir, H.; Saemundsson, K.; Stefánsson, V.; Steingrímsson, B.; Tómasson, J.; Kristjánsson, L. The Iceland crust: Evidence from drillhole data on structure and processes. *Deep Drill. Results Atl. Ocean: Ocean Crust* **2011**, *2*, 43–65.

54. Odin, G. Introduction to the Study of Green Marine Clays. *Dev. Sedimentol.* **1988**, *45*, 1–3.

55. Beyssac, O.; Rouzaud, J.-N.; Goffé, B.; Brunet, F.; Chopin, C. Graphitization in a high-pressure, low-temperature metamorphic gradient: a Raman microspectroscopy and HRTEM study. *Contrib. Miner. Pet.* **2002**, *143*, 19–31. [CrossRef]

56. Beyssac, O.; Goffé, B.; Chopin, C.; Rouzaud, J.N. Raman spectra of carbonaceous material in metasediments: a new geothermometer. *J. Metamorph. Geol.* **2002**, *20*, 859–871. [CrossRef]

57. Huang, E.-P.; Huang, E.; Yu, S.-C.; Chen, Y.-H.; Lee, J.-S.; Fang, J.-N. In situ Raman spectroscopy on kerogen at high temperatures and high pressures. *Phys. Chem. Miner.* **2010**, *37*, 593–600. [CrossRef]

58. Pimenta, M.A.; Dresselhaus, G.; Dresselhaus, M.S.; Cançado, L.G.; Jorio, A.; Saito, R. Studying disorder in graphite-based systems by Raman spectroscopy. *Phys. Chem. Chem. Phys.* **2007**, *9*, 1276–1290. [CrossRef]

59. Maheshwari, R.; Bharadwaj, G.; Bhat, M.K. Thermophilic Fungi: Their Physiology and Enzymes. *Microbiol. Mol. Biol. Rev.* **2003**, *64*, 461–488. [CrossRef]

60. Ivarsson, M.; Broman, C.; Lindblom, S.; Holm, N. Fluid inclusions as a tool to constrain the preservation conditions of sub-seafloor cryptoendoliths. *Planet. Space Sci.* **2009**, *57*, 477–490. [CrossRef]

61. Stakes, D.S.; Scheidegger, K.F. Temporal variations in secondary minerals from Nazca plate basalts, diabases, and microgabbros. *From Rodinia to Pangea: The Lithotectonic Record of the Appalachian Region* **1981**, *154*, 109–130.

62. Alt, J.C.; Honnorez, J. Alteration of the upper oceanic crust, DSDP site 417: mineralogy and chemistry. *Contrib. Miner. Pet.* **1984**, *87*, 149–169. [CrossRef]

63. Robb, L. *Introduction to Ore-Forming Processes*; John Wiley & Sons: Hoboken, NJ, USA, 2005; ISBN 0632063785.

64. Henin, S. Synthesis of Clay Minerals at Low Temperatures. *Clays Clay Miner.* **2006**, *4*, 54–60. [CrossRef]

65. Mariner, R.H.; Surdam, R.C. Alkalinity and Formation of Zeolites in Saline Alkaline Lakes. *Science* **2006**, *170*, 977–980. [CrossRef]

66. Savage, D.; Rochelle, C.; Moore, Y.; Milodowski, A.; Bateman, K.; Bailey, D.; Mihara, M. Analcime reactions at 25–90 °C in hyperalkaline fluids. *Miner. Mag.* **2001**, *65*, 571–587. [CrossRef]

67. Gudbrandsson, S.; Wolff-Boenisch, D.; Gislason, S.R.; Oelkers, E.H. An experimental study of crystalline basalt dissolution from $2 \leq pH \leq 11$ and temperatures from 5 to 75 °C. *Geochim. Cosmochim. Acta* **2011**, *75*, 5496–5509. [CrossRef]

68. Rousk, J.; Brookes, P.C.; Bååth, E. Contrasting Soil pH Effects on Fungal and Bacterial Growth Suggest Functional Redundancy in Carbon Mineralization. *Appl. Environ. Microbiol.* **2009**, *75*, 1589–1596. [CrossRef]

69. Reitner, J.; Schumann, G.; Pedersen, K. Fungi in subterranean environments. In *Fungi in Biogeochemical Cycles*; Cambridge University Press: Cambridge, UK, 2006; p. 377. ISBN 9780511550522.

70. Benson, S.A. Chapter 2 The bacterial outer membrane and surface permeability. *Dev. Biol.* **1998**, *9*, 15–22.

71. Rodríguez-Navarro, A. Potassium transport in fungi and plants. *Biochim. Biophys. Acta* **2000**, *1469*, 1–30. [CrossRef]

72. Gadd, G.M. Geomycology: biogeochemical transformations of rocks, minerals, metals and radionuclides by fungi, bioweathering and bioremediation. *Mycol. Res.* **2007**, *111*, 3–49. [CrossRef]

73. Gadd, G.M. Metals, minerals and microbes: Geomicrobiology and bioremediation. *Microbiology* **2010**, *156*, 609–643. [CrossRef]

74. Vaughan, D.J.; Pattrick, R.A.D.; Wogelius, R.A. Minerals, metals and molecules: ore and environmental mineralogy in the new millennium. *Miner. Mag.* **2002**, *66*, 653–676. [CrossRef]

75. Brown, G.E.; Trainor, T.P.; Chaka, A.M. Geochemistry of Mineral Surfaces and Factors Affecting Their Chemical Reactivity. In *Chemical Bonding at Surfaces and Interfaces*; Elsevier: Amsterdam, The Netherlands, 2008; pp. 457–509.

76. Bernal, J.D. *The Physical Basis of Life*; Routledge and Paul: London, UK, 1951; p. 597.

77. Clarkson, E. Genetic Takeover and the Mineral Origins of Life. By A. G. Cairns-Smith (first published 1982, first paperback edition 1987). Cambridge University Press. ISBN 0 521 34682 7. *Genet. Res.* **1988**, *51*, 167–168. [CrossRef]

78. Hashizume, H. Role of Clay Minerals in Chemical Evolution and the Origins of Life. In *Clay Minerals in Nature—Their Characterization, Modification and Application*; IntechOpen: London, UK, 2012.

79. Smith, S.; Read, D. *Mycorrhizal Symbiosis*, 3rd ed.; Academic Press: Cambridge, MA, USA, 2008; ISBN 9780123705266.

Article

Microbial Diversity in Sub-Seafloor Sediments from the Costa Rica Margin

Amanda Martino [1,2,*], Matthew E. Rhodes [3], Rosa León-Zayas [4,5], Isabella E. Valente [3], Jennifer F. Biddle [5] and Christopher H. House [2]

[1] Department of Biology, Saint Francis University, Loretto, PA 15940, USA
[2] Department of Geosciences, Penn State University, University Park, PA 16802, USA; chrishouse@psu.edu
[3] Department of Biology, College of Charleston, Charleston, SC 29424, USA; rhodesme@cofc.edu (M.E.R.); valentebe@g.cofc.edu (I.E.V.)
[4] Department of Biology, Willamette University, Salem, OR 97301, USA; rleonzayas@willamette.edu
[5] School of Marine Science and Policy, University of Delaware, Lewes, DE 19958, USA; jfbiddle@udel.edu
* Correspondence: amartino@francis.edu

Received: 31 March 2019; Accepted: 11 May 2019; Published: 13 May 2019

Abstract: The exploration of the deep biosphere continues to reveal a great diversity of microorganisms, many of which remain poorly understood. This study provides a first look at the microbial community composition of the Costa Rica Margin sub-seafloor from two sites on the upper plate of the subduction zone, between the Cocos and Caribbean plates. Despite being in close geographical proximity, with similar lithologies, both sites show distinctions in the relative abundance of the archaeal domain and major microbial phyla, assessed using a pair of universal primers and supported by the sequencing of six metagenomes. *Elusimicrobia, Chloroflexi, Aerophobetes, Actinobacteria, Lokiarchaeota,* and *Atribacteria* were dominant phyla at Site 1378, and *Bathyarchaeota, Chloroflexi, Hadesarchaeota, Aerophobetes, Elusimicrobia,* and *Lokiarchaeota* were dominant at Site 1379. Correlations of microbial taxa with geochemistry were examined and notable relationships were seen with ammonia, sulfate, and depth. With deep sediments, there is always a concern that drilling technologies impact analyses due to contamination of the sediments via drilling fluid. Here, we use analysis of the drilling fluid in conjunction with the sediment analysis, to assess the level of contamination and remove any problematic sequences. In the majority of samples, we find the level of drilling fluid contamination, negligible.

Keywords: subsurface; sediment; bacteria; archaea; deep biosphere

---

1. Introduction

Scientific drilling programs (Deep Sea Drilling Program-DSDP, Ocean Drilling Program-ODP, Integrated Ocean Drilling Program-IODP, and the International Ocean Discovery Program-IODP) have enabled successful studies of microbiology in the deep marine sub-surface. The Peru Margin, Canterbury Basin, Nankai Trough, and Cascadia Margin are just a few of the areas of sediment drilled in order to examine the deep biosphere [1–4]. However, many areas of ocean sediment remain to be explored. This study provides the first information on microbial community composition and potential biogeochemical relationships present at one of these previously unexplored locations. The Costa Rica Margin sub-seafloor has been targeted by the larger IODP community, due to the unique properties and history of the erosive subduction zone, between the Cocos and Caribbean plates [5]. The Costa Rica subduction system is of interest as it may be a tangible way to see the connection between geology and biology, as influenced by subductive processes [6]. For the deep biosphere, this location is a useful comparison to those that have been investigated previously in other coastal margin locations (for example: Peru, Nankai, Cascadia margin). While some previous margin studies have shown archaea to constitute only a very minor portion of the microbial population of the seafloor [3,7,8], others have suggested a greater influence [9,10]. The environmental conditions in the sub-seafloor that constrain archaeal abundance are currently unknown, and the dominant microbial populations of the Costa Rica Margin are also unknown.

In addition to examining the microbial community composition, this paper also examines the influence of drilling fluid contamination in deep sediment cores. While numerous sub-seafloor sediment microbial communities have now been examined, much of the research has been focused on the first 100–200 m of the sediment column, as this is generally the depth at which advanced piston coring (APC) is successful. Beyond these depths, the sediment column becomes more compacted, requiring extended core barrel (XCB) drilling. Contamination control protocols, developed during ODP Leg 185, and applied during most of the microbiological sampling thereafter, have shown that a substantial increase in penetration of contaminating drilling fluid into the sediment core occurs after the switchover to XCB coring [11–13]. Since contaminating microbes may mask or out-compete indigenous subsurface life, current practices focus on reducing the potential for contamination introduction, which has limited most microbiological research to APC-retrieved samples. While it may not be possible, with current drilling methods, to entirely eliminate the contamination of samples during XCB drilling, several studies suggest that the development of methods to investigate indigenous subsurface microbes, despite the presence of contaminating microbes, is worth pursuing [14,15]. Identifying the methods of handling the potential contamination of XCB coring would allow an increased usage of deeper samples for studies of the deep biosphere. Hence, another important feature of the present study, is the utilization of XCB-drilled sub-seafloor sediment cores, through the concurrent molecular analysis of both the sediment samples of interest, and the drilling fluid used in obtaining those samples.

Initial DNA-based studies of the sub-seafloor microbial community composition utilized PCR reactions, targeting the 16S rRNA gene, followed by the construction of clone libraries, in order to amplify DNA fragments prior to Sanger sequencing, from which one would generally obtain several hundred DNA sequences to analyze, as reviewed in [16,17]. In contrast, next-generation DNA sequencing methods, not only eliminate the need for extra steps involved in clone library creation, and thus reduce bias, but also reduce the time and cost of sample preparation. These high throughput sequencing methods allow a more confident assessment of microbial community composition, including the ability to analyze in-situ samples with different measures of contamination [18]. Given the nature of the sampling process used for sub-seafloor microbiology, it is important to have a solid understanding of the potential influence of drilling fluid contamination on molecular-based community composition and diversity analyses. While few studies have looked at the composition of drilling fluid microbial communities [14,19], there is still need for a comprehensive examination of the potential influence of those sequences on in-situ community analyses. A direct comparison of the lineages found in the sediment samples to those found in the drilling fluid, used in obtaining those samples, from both advanced piston coring (APC) and extended core barrel (XCB) methods, would allow for isolation of the in-situ signal. This type of analysis was recently produced for samples from the R/V Chikyu drilled by different methods, and the successful identification and removal of potential contaminants was demonstrated [18].

In the following, we present the first microbial community analysis of the Costa Rica Margin sub-seafloor environment, using 16 sediment samples obtained during IODP Expedition 334. We provide a thorough comparison of the community composition of the sub-seafloor sediment samples to that of a sample of the drilling fluid used to obtain the samples during coring operations. We used a 16S rRNA primer set, designed to target both domains for high throughput PCR-based sequencing. In order to provide some means of analyzing the accuracy of the primer set, as all PCR primers are subject to bias, we include further testing results of the primer set used on several environmental samples where the expected archaeal: Bacterial ratio is well known, and supplement PCR results with those of metagenomic sequencing, which does not utilize specific primers. We determine shifts in the community composition, with depth and with sampling site, and assess correlations between microbial taxa and environmental chemistry.

2. Materials and Methods

2.1. Sample Collection

Sediment samples were collected in March and April of 2011, from sites 1378 and 1379 of IODP Expedition 334, Costa Rica Seismogenesis Project (CRISP: Figure 1, Table S1). Detailed site descriptions can be found in the IODP Proceedings for Expedition 334 [5], but those descriptions are summarized as follows: Both sites were located along the Costa Rica Margin, on the upper plate of a convergent plate boundary between the Cocos plate and the Caribbean plate (part of the larger Middle America Trench system). At Site U1378, located on the middle slope region 38 km offshore of the Osa Peninsula of Costa Rica, sediment extended ~750 m to the upper-plate basement. Coring was successful down to a total of 523 m below seafloor (mbsf), at which point hole conditions prevented further drilling, with the first 128 m taken with the APC coring system before switching to the XCB system [5]. At Site U1379, located on the upper slope region 34 km offshore of Osa Peninsula, sediment extended ~890 m to basement. Coring was successful all the way through the sediment and into the upper basement, with the switchover from the APC to XCB system, occurring at 91 mbsf. The sediments at both sites were described as being dominantly silty clay to clay, with interspersed sandy layers (cm scale for Site 1378 and decimeter scale for 1379) [5]. At site 1378, tephra layers were observed between 0 and 128 mbsf, whereas tephras were located below 177 mbsf at Site 1379. Cell counts performed in the upper 250 m of sediments from Site 1378 and the upper 215 m of sediment from Site 1379 ranged from 10^6 and 10^8 cells per cubic centimeter and generally declined with depth [5].

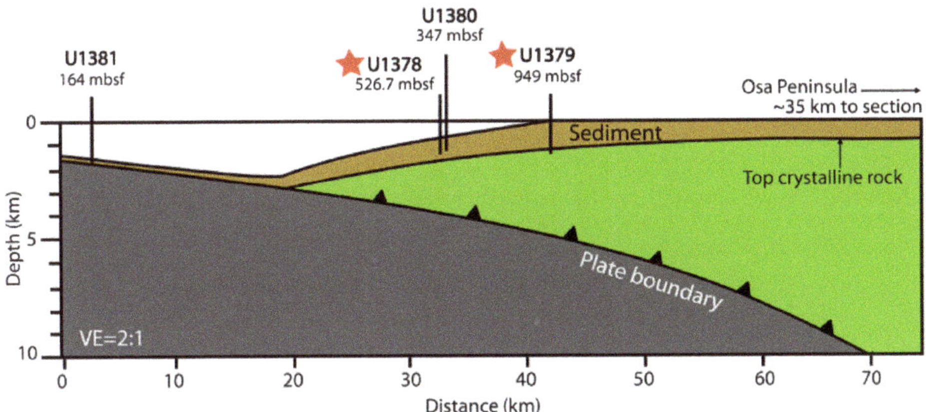

Figure 1. Wide-angle seismic section showing locations and depths of sites drilled during International Ocean Discovery Program (IODP) Expedition 334. Stars indicate sites utilized in current study [5,20].

Microbiology samples were obtained at intervals down to 500 mbsf at 1378 and 780 mbsf at 1379. The microbiology samples, retrieved during this expedition, were always taken from the section of sediment core immediately adjacent to samples used for pore-water geochemistry analyses. Further, samples were taken with great care to reduce any possible sources of outside contamination. Whole round sediment core samples were cut from the intact core sections on deck with sterile instruments. These whole round cores were subsampled with modified sterile syringes (luer-lock end removed to allow large opening) after the removal of the upper few millimeters of sediment from the core surface. Subsamples were then immediately placed into −80 °C freezers and were kept frozen until subsequent DNA extraction. Additionally, a sample of the drilling fluid, used during operations on this cruise, was taken directly from the supply in-use during drilling at site 1378 and stored at −80 °C until later DNA extraction.

Three additional samples were utilized for the purpose of supplementary testing of our 16S rRNA universal primer set. The Dead Sea water sample was obtained in 2007, and details have been previously described [21]. The Pacific Ocean seawater sample was collected during IODP Expedition 334, from sampling surface water while stationed at Site 1379 on 25 March 2011. Lastly, the anaerobic digester effluent sample was extracted from a lab-scale fixed-film anaerobic digester system located at Penn State University Department of Geosciences.

2.2. DNA Extraction

PCR-amplifiable DNA was successfully extracted from 9 samples from site 1378, 9 samples from site 1379, and the sample of drilling fluid. DNA was extracted using the PowerSoil DNA Isolation Kit (MO-BIO Laboratories, INC., USA) according to the manufacturer's protocol with modifications as follows. For the sediment samples, 4 aliquots were used per sample, each 0.3–0.4 g. For the drilling fluid, 4 aliquots of 200 µL each were used. After the addition of solution C1 in the beginning of the protocol, the samples were incubated in a 65 °C water bath for 15 min. Following this step, for each sample, 2 of the replicates received 20 s. of bead beating and 2 received 30 s, and all 4 were combined during the elution step of the extraction, to concentrate retrieved DNA. The extractions for each sample were carried out independently to avoid cross-contamination, and each was done with a blank as well, to which no sample was added, but on which all steps of the procedure were carried out. A representative extraction blank was selected for sequencing as a quality control measure (though all blanks were checked with PCR and no bands were visible). Additionally, all extractions were performed in a laminar flow hood located in a PCR-free laboratory.

Extraction of DNA from the 3 supplementary samples used for primer validation was as follows: 1. Dead Sea sample: DNA extraction was carried out previously as described [21]; 2. Pacific Ocean seawater sample: Approximately 235 mL of the seawater sample was passed through a Sterivex-GV Durapore 0.22 µm filter (EMD Millipore Corp.), from which the filter was then removed and placed into a bead tube from the FastDNA Spin Kit for Soil (MP Biomedicals). DNA was then extracted following the manufacturer's protocol. At the final step, DNA was eluted to a volume of 100 µL; 3. Anaerobic digester effluent sample: DNA was isolated, using a Powersoil kit (MoBio), according to the manufacturer's instructions, with 40 s of bead beating.

2.3. 16S rRNA PCR and Amplicon Sequencing

For 16S rRNA analyses, a fragment of the gene encompassing the V6–V9 hypervariable regions was targeted with the primer pair 926F (5′- AAACTYAAAKGAATTGRCGG-3′) and a modified version of 1392R (5′-ACGGGCGGTGTGTRC-3′), developed previously for amplification of DNA from oil sands [21]. The primers were modified further with the addition of 454 Life Science's A or B sequencing adapters, as well as multiplex identifier (MID) tags permitting multiple samples to be sequenced together. PCR reactions were set up utilizing illustra PuReTaq Ready-To-Go PCR beads (GE HealthCare), with 10 µM of each the forward and reverse primers, and 2 µL of extracted DNA in a total volume of 25 µL. PCR cycling conditions included an initial 5 minute denaturing at 94 °C, followed by 32 cycles (for Costa Rica sediment samples) of 1 min 94 °C, 30 sec at 53 °C, and 2 min at 72 °C, followed by a final 10 minute extension at 72 °C. For the Pacific Ocean seawater sample, only 25 cycles were needed, and for the anaerobic digester sample, 4 separate PCR reactions were run with 15, 20, 25, and 30 cycles, in order to analyze possible effects of cycle number on community composition analysis. PCR cycling conditions for the Dead Sea sample were as described [21]. All PCR products were gel-purified using a PrepEase Gel Extraction Kit (Affymetrix, Inc.) according to the manufacturer's instructions. Sequencing was carried out at the Pennsylvania State University on a Roche 454 Genome Sequencer FLX+ sequencing system (454 Life Sciences) using Titanium chemistry. Amplicons were sequenced in one direction only.

2.4. 16S rRNA Amplicon Data Analysis

Sample demultiplexing was performed using the mothur package [22], as well as some preliminary quality controls, eliminating sequences shorter than 100 bp, with more than one mismatch in their barcode sequence, or with more than 2 mismatches in their primer sequence. In addition, sequences were screened by quality score using the "qwindowaverage" function in mothur, set at a quality of 35. Resulting individual fasta files were then processed with the NGS analysis pipeline of the SILVA rRNA gene database project (SILVAngs 1.0) [23]. This pipeline included alignment with the SILVA Incremental Aligner (SINA v 1.2.10 for ARB SVN (revision 21008)) [24] against the SILVA SSU rRNA SEED as well as quality control steps [23]. Specifically, reads that had fewer than 50 aligned nucleotides and/or more than 2% ambiguities or homopolymers were excluded from further analysis, as well as putative contaminants, artefacts, and reads with low alignment quality. The remaining sequences were de-replicated and clustered into OTUs on a per sample basis, using cd-hit-est (version 3.1.2) [25] running in accurate mode, ignoring overhangs, and applying identity criteria of 1.00, and 0.98, respectively. A reference sequence from each OTU was classified using a local nucleotide BLAST search against the non-redundant version of the SILVA SSU Ref dataset (release 115; [26]) using blastn (version 2.2.28+; [27]) with standard settings [28]. Because of some of the limitations in BLAST, classification was only assigned to a read if the value of the function (%sequence identity+% alignment coverage)/2 was greater than 93. Unclassified reads were assigned to the group "No Relative." Classifications of the reference sequences were then mapped onto all reads within their respective OTUs. Additional information on the processing of sequencing blanks is available in the supplemental material.

2.5. Metagenomic Sequencing and Analysis

Five of the six Costa Rica sediment samples were sequenced on an Illumina HiSeq, while one was sequenced with 454 Life Sciences pyrosequencing technology. For the 3 samples from Site 1378, extracted DNA was sent to the Marine Biology Laboratory (Woods Hole, MA, USA) had libraries constructed and were sequenced on an Illumina Hiseq 1000. Libraries were constructed to be ~170bp long molecules with ~25–30 bp partial overlap between pairs of reads. Each sample received 1/6th of a lane of sequencing. The sample from 2.88 mbsf had 79,382,253 sequences, 32.29 mbsf had 33,398,960 sequences and 93.98 mbsf had 31,536,056 sequences. For 2 samples from site 1379 (22 mbsf and 45 mbsf) extracted DNA had libraries were constructed with a 350 bp insert and was sequenced by the Schuster lab at Penn State University on an Illumina HiSeq 1000. Each sample received about 1/9th of a lane of sequencing. After removal of adapter sequences, and due to quality issues the first 15 nucleotides and any sequences with ambiguous bases, the sample from 22 mbsf had 118,990,546 sequences, and that from 45 mbsf had 19,804,126 sequences. One additional sample from 1379 (3 mbsf) was sequenced with a different method as it was originally sequenced for a different study. This sample was also sequenced by the Schuster lab, but on a Roche 454 Genome Sequencer FLX sequencing system (454 Life Sciences) with Titanium chemistry. The sample had 300,922 sequences with an average length of 529 bp. Metagenomic sequencing information for primer validation samples is provided in the supplementary material.

Metagenomes, resulting from the 454 sequencing platform, were quality controlled using the prinseq program [29]. Parameters were set to remove any sequence shorter than 60 bp, longer than 800 bp, with an average quality score less than 20, with greater than one ambiguous base, and all replicates. Metagenomes resulting from Illumina sequencing were quality controlled using the trimmomatic program [30] and default parameters. All metagenomes reads were subsequently assigned to taxa using the Kaiju web server [31]. Reads were annotated against the nr + euk database using the greedy run mode with a minimum match length of 11, minimum match score of 75, and five allowed mismatches. Metagenomes were also classified using marker genes by the PhyloSift program using default parameters [32]. In order to standardize analysis across sequencing platforms, we found that using 1 million of the shorter, paired-end Illumina reads, resulted in a similar percentage of protein coding genes being detected as the full dataset (300K reads) of the longer 454 reads.

2.6. Correlations

Geochemical data was obtained from the IODP data repository, with refinements from a shipboard geochemist ([5]; E. Solomon, pers. comm.; (Br Data) [33]; (Sr, B, and Li Data) [34]). The software package SPSS Statistics (IBM) was used to run bivariate correlations between microbial phyla and geochemical parameters, using the default settings for this function. For each phylum, all geochemical correlations were run simultaneously, allowing for the calculation of significance values taking into account the total number of variables tested. These tests produced Pearson Correlation Coefficients and associated significance values (from a 2-tailed test) for each pair of variables tested.

2.7. Accession Numbers

All new amplicon and metagenomic datasets for this study have been deposited in the EMBL-EBI European Nucleotide Archive (ENA) under the study accession number PRJEB11766. The metagenomic and amplicon datasets from the Dead Sea samples were obtained from previously published studies [21,35].

3. Results and Discussion

3.1. Microbial Community Analysis Reveals High Proportion Archaea

Eight samples from Site 1378, and seven samples from Site 1379, were used to examine the microbial community composition of the Costa Rica Margin sub-seafloor environment by amplicon sequencing using universal primers. Examining the results of the 16S rRNA data analysis, we found a significant proportion of archaea represented at both sites (Figure 2, Panel A). For Site 1378, the proportion of total classified amplicon sequences identified as Archaea was 58.4% for the shallowest sample examined (2.88 mbsf). The proportion of archaea then declined to 23.7% at 32.29 mbsf, which is no longer greater than bacteria, but a significant portion of the population remaining through 158.24 mbsf. The lowest proportion was then observed at the greatest depth, with only 1.3% of sequences from the 329.1 mbsf sample. At Site 1379, sequences identified as archaea made up 60.9% of classified amplicon sequences for the shallowest sample analyzed (2.90 mbsf), out-numbering bacterial sequences from that depth. However at this site, archaea remained dominant with depth, with more archaeal sequences than bacterial sequences at 3 additional depths: 67.28 mbsf (60.4%), 89.38 mbsf (66.9%), and 115.07 mbsf (71.7%). Of the 3 depths where this was not the case, the lowest that the archaeal proportion dropped was to 26.0% of sequences identified, which occurred in the deepest sample at 211.53 mbsf. It should also be noted that the greatest proportion of archaea at Site 1379 occurred not in the shallowest sample, but rather at 115.07 mbsf (71.7%).

One of the major differences in these sites is sedimentation rate, which would potentially vary the composition of organic matter at depth [5]. More archaea may be at Site 1379, due to its nearly doubled sedimentation rate compared to Site 1378, which suggests more readily available organic matter is delivered to depth. Previous margin work suggests the majority of subsurface archaea are heterotrophic and would be reliant on this organic matter [9,10]. However, detailed measurements on the state of organic matter are not available from this site and beyond the scope of the current study, leaving us with this testable hypothesis for the future.

While the primers were not designed to target Eukaryotes, some matches occurred, and were left in the domain-level analysis, in order to show that there is evidence of a Eukaryotic signature in some locations at these sites (Figure 2; Panel A). Because the primers are not optimized for Eukaryotes, however, it should be cautioned that these data likely do not provide an accurate representation of the abundance or diversity of Eukaryotes. In the domain-level analysis presented, the matches to Eukaryote sequences represented much less than 1% of the total classified sequences for most samples, with exceptions at Site 1378 of 2.3% identified at 158.24 mbsf and 9.1% identified at 329.1 mbsf.

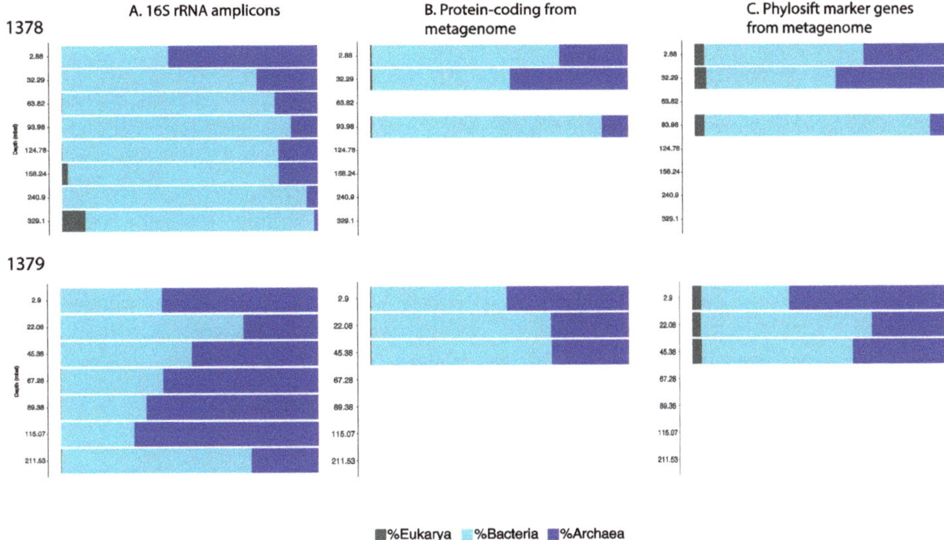

Figure 2. Relative proportions of Bacteria, Archaea and Eukaryotes for Sites 1378 and 1379 assessed using 3 methods. Shown are the percentages of total classified reads for each method. A. 16S rRNA amplicons produced via PCR with universal primers were classified with the SILVAngs pipeline. B. Protein-coding genes from metagenomic sequences were classified using Kaiju. C. A set of phylogenetic marker genes from the metagenomic sequences were classified using Phylosift.

Due to the known bias created by PCR primers, we endeavored to assess the utility of our primer set as "universal" 16S rRNA primers through several confirmation tests. First, we used a second method of analysis on a selection of our Costa Rica Samples to see if any significant discrepancies were observed. We performed metagenomic sequencing on 3 samples from Site 1378 and 3 samples from Site 1379. Metagenomic sequencing does not rely on specific PCR primers, and so bias from primer specificity is not a factor, although other biases still exist. The 6 samples used for metagenomic sequencing were chosen with the aim of getting representation at multiple depths, but limited by the fact that our metagenomic sequencing procedure required a higher concentration of DNA for success, and many of our deeper samples did not produce enough DNA without using whole genome amplification, which would have further biased the results.

The metagenomic sequences were classified using two different methods, again for purposes of comparison, in this case due to possible database bias. The first method was to obtain taxonomic information from all protein-coding sequences using the program Kaiju, which ranged from 8 to 16% of the total data [31] (Figure 2, Panel B), and the second was to use only single-copy phylogenetically informative marker genes using the PhyloSift platform [32] (Figure 2, Panel C). Both methods of metagenomic sequence analysis of the selected samples corroborated the idea that archaea are a significant portion of the microbial community at these two sub-seafloor locations. The proportions of archaea to bacteria differed in some samples compared to the amplicon analysis however, so the true trend of archaeal abundance with depth at these sites is difficult to determine from these data.

In addition to the metagenome to amplicon comparison of the Costa Rica sub-seafloor samples, we also assessed the primer set using a similar comparison on three well-characterized environments not directly related to this study, with known, and differing, archaeal proportions. Sequence results from both methods again (amplicon sequencing and metagenomic sequencing) were obtained from a water sample of the Dead Sea (archaeal dominated environment), a Pacific Ocean surface seawater sample (a bacterial dominated environment), and an effluent sample from a laboratory microbial anaerobic digester (bacteria dominated but with a significant number of archaea as well). In all cases, the archaea-to-bacteria ratio remained highly consistent (Figure 3). For the archaeal-dominated Dead Sea, the archaea comprised 95.3% of protein-coding reads from the metagenome, 92.2% of the marker genes from the metagenome, and 98.4% of the amplicon dataset. For the bacterial-dominated sample, Pacific Ocean seawater, archaea comprised 2.9% of the protein-coding reads from the metagenome, 11.8% of the marker genes from the metagenome, and 10.2% of the amplicon dataset. For our second bacterial-dominated sample, anaerobic digester effluent, archaea made up 16.4% of the protein-coding reads from the metagenome, 16.9% of the marker genes from the metagenome, and 14.4% of the amplicon dataset, measured after 25 PCR cycles. For this sample, we also examined the possible changes in bias, due to number of PCR cycles, and thus have data representing the composition at 15, 20, 25, and 30 PCR cycles (Figure 3). While this analysis shows an increase in proportion of archaea with cycle number (6.4%, 8.2%, 14.4%, 19.1%, respectively), that proportion remains well below the proportion of bacteria and still very close to results from the metagenomic analysis. These results show that in these three separate environments, the primer set successfully approximated the expected archaeal ratio when compared to that obtained from metagenomic analysis, and when compared to the ratio expected for these domains in these well-studied environments.

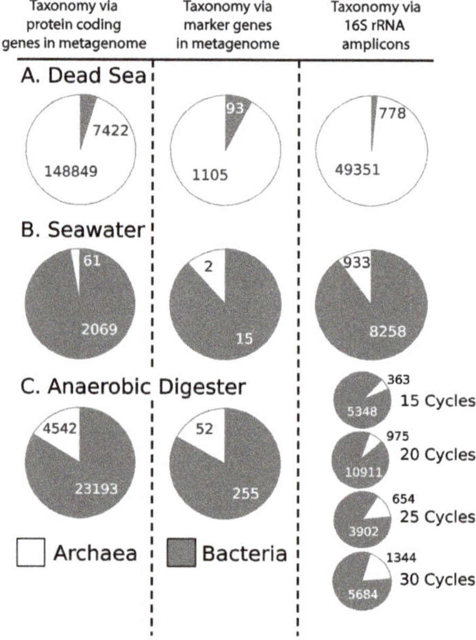

Figure 3. Taxonomy calls based on protein coding genes from metagenome (left), phylogeny from marker genes in metagenomes (center) and 16S rRNA amplicon reads using universal primers (right) on additional samples to show universal primer effectiveness. (**A**) Data from the Dead Sea; (**B**) Data from seawater; (**C**) Data from an anaerobic digester, with additional investigations into the robustness of PCR cycles for amplicon data. Archaea are indicated by white and bacteria are indicated by gray, all samples shown as percentage abundances.

3.2. Taxonomic Analysis Beyond the Domain Level

At the phylum level, when considering each site as a whole (all depths examined together), the dominant phyla identified were *Elusimicrobia*, *Chloroflexi*, *Aerophobetes*, *Actinobacteria*, *Lokiarchaeota*, and *Atribacteria* for Site 1378, and *Bathyarchaeota*, *Chloroflexi*, *Hadesarchaeota*, *Aerophobetes*, *Elusimicrobia*, and *Lokiarchaeota* for Site 1379 (Figure 4). One of the most notable differences in phylum-level composition between the two sites, is the difference in abundance of *Bathyarchaeota*, being much more abundant at Site 1379. Examining the *Bathyarchaeota* (formerly known as Miscellaneous Crenarchaeotal

Group, MCG) with depth at each site (Table 1), the phylum is present in a high proportion, at both sites, at the most shallow depth examined, 30.54% for Site 1378, and 34.81% for 1379. However, while that proportion remains high and even goes higher at times with depth at Site 1379, it is fairly low at all other depths of Site 1378. *Bathyarchaeota* have been found to be common and widespread in the sub-seafloor environment, and their metabolic capabilities are very broad, including acetogenesis, methane metabolism, dissimilatory nitrogen and sulfur reduction, and others [36]. In this taxonomic classification, bathyarchaeal sequences were not able to be further sub-classified beyond the phylum level.

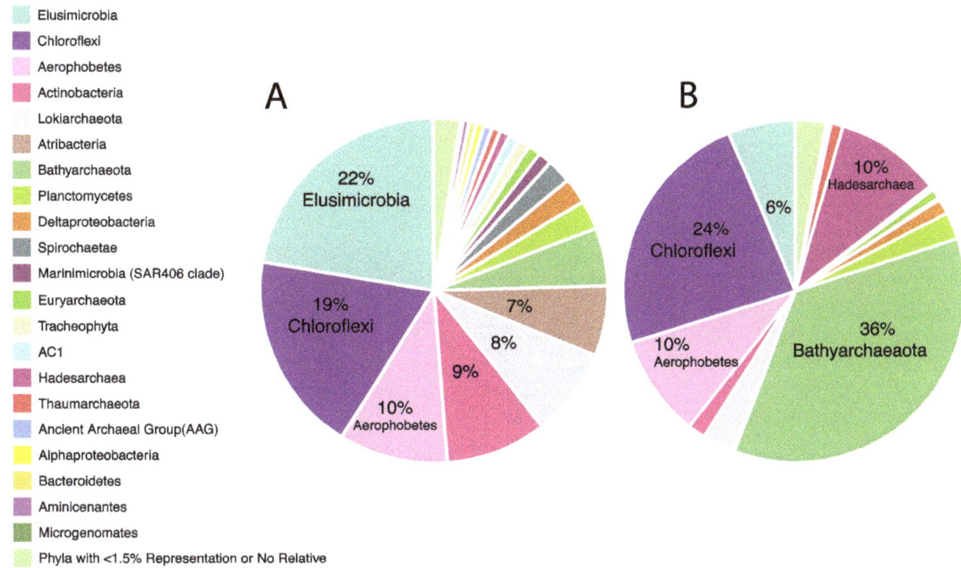

Figure 4. Detailed taxonomic breakdown of overall 16S rRNA amplicon sequences at (**A**) Site 1378 and (**B**) Site 1379. Data shown by percentage abundance.

When examining which phyla appeared less often in 1379 as compared with 1378, the most notable is the *Elusimicrobia*, so it may be that *Bathyarchaeota* is occupying a niche space of *Elusimicrobia*, at Site 1379, and vice versa. Previously, *Bathyarchaeota* have been described as anaerobic protein degrading microbes [36], which is similar to the functions given to *Elusimicrobia*, determined in the termite gut [37]. Examining the trend with depth, at Site 1378 *Elusimicrobia* are present in low abundance at 2.88 mbsf and 32.39 mbsf, but then are prominent over the next 4 depths examined, from 63.82 mbsf to 158.24 mbsf, dipping some at 240.90 mbsf before soaring to 54.81% at 329.1 mbsf. At Site 1379, *Elusimicrobia* reach notable proportions at only 2 depths, 45.38 mbsf and 211.53 mbsf, but are still present throughout all samples (Table 1). Other notable taxa that are present in lower abundance at Site 1379 and therefore may be influenced by the dominance of *Bathyarchaeota* include *Actinobacteria*, *Atribacteria* and *Lokiarchaeota*. Also notable, at Site 1378 when *Elusimicrobia* dips to only 5% of the sequence reads at 240.90 mbsf, *Actinobacteria* rises to 44.58% of all sequences for that depth, its highest prominence by far. For that 1378 sample, the rise in *Actinobacteria* was nearly all due to sequences classified as group "OPB41," which was also the most common *Actinobacteria* in numerous other samples as well.

Table 1. Detailed taxonomic percent abundances by depth of 16S rRNA amplicon sequencing.

						1378							1379		
Sample Depth (mbsf):	2.88	32.29	63.82	93.98	124.78	158.24	240.90	329.10	2.90	22.08	45.38	67.28	89.38	115.07	211.53
Archaea:															
Bathyarchaeota	30.54	1.88	0.00	0.88	4.29	3.66	2.08	0.24	34.81	18.21	28.57	50.86	51.59	46.30	20.04
Lokiarchaeota	16.65	14.46	14.29	3.46	8.22	6.00	1.76	1.05	7.51	3.81	5.27	0.46	0.04	0.01	3.39
Thaumarchaeota	4.00	0.91	0.48	0.21	0.57	0.33	0.02	0.00	1.77	1.74	1.03	1.59	1.14	0.69	0.14
Euryarchaeota	1.94	0.78	0.00	4.64	1.18	1.01	0.03	0.00	2.54	1.04	2.23	0.32	0.37	0.01	0.00
Hadesarchaea	1.34	1.14	0.12	0.00	0.52	4.16	0.16	0.00	11.30	3.69	11.56	5.53	12.84	24.23	1.22
Ancient Archaeal Group (AAG)	0.04	4.23	1.71	0.00	0.27	0.00	0.00	0.00	0.00	0.00	0.00	0.00	0.00	0.00	0.00
Bacteria:															
Chloroflexi	20.61	27.29	13.19	19.83	14.52	25.30	23.64	5.52	22.18	47.48	22.30	28.00	16.93	13.09	14.71
Planctomycetes	5.83	3.53	1.48	0.89	1.43	4.71	3.74	0.00	6.00	2.96	2.30	1.87	1.36	1.40	2.05
Deltaproteobacteria	5.07	3.56	3.59	4.25	0.82	1.32	0.38	0.12	2.53	2.69	1.61	0.61	0.09	0.06	2.31
Aminicenantes	2.09	0.77	1.61	0.10	0.00	0.26	0.01	0.00	0.34	0.23	0.37	0.50	0.19	0.13	0.10
Microgenomates	1.63	0.00	0.00	0.00	0.00	0.00	0.00	0.00	1.01	0.00	0.00	0.11	0.00	0.00	0.00
Aerophobetes	1.32	10.35	14.72	20.05	10.66	19.08	4.46	0.38	2.40	4.18	5.06	2.56	8.68	7.14	36.96
AC1	0.68	2.13	2.24	1.75	0.85	0.26	0.06	0.00	0.14	0.35	0.75	0.92	0.13	0.03	0.39
Spirochaetae	0.64	0.61	2.37	0.90	1.20	0.61	0.44	11.56	0.23	0.20	0.27	0.00	0.00	0.00	0.00
Elusimicrobia	0.40	3.07	23.65	29.72	42.60	19.16	5.03	54.81	1.96	1.95	14.66	1.75	2.89	4.99	15.98
Atribacteria	0.37	9.34	3.65	8.39	8.12	3.10	9.35	10.22	0.08	0.28	0.07	0.12	0.04	0.03	0.53
Marinimicrobia (SAR406 clade)	0.32	2.17	2.70	0.58	0.38	0.91	1.58	1.61	0.13	0.08	0.32	0.08	0.03	0.05	0.00
Actinobacteria	0.23	10.17	12.53	0.51	2.17	4.94	44.58	0.51	0.20	7.85	1.44	0.40	0.59	0.93	0.00
Bacteroidetes	0.05	1.60	1.09	0.12	0.34	0.23	0.71	0.93	0.05	0.02	0.08	0.02	0.00	0.02	0.06
Alphaproteobacteria	0.00	0.00	0.00	0.06	0.06	1.97	0.57	3.00	0.00	0.00	0.00	0.00	0.01	0.09	0.00
Eukaryota:															
Tracheophyta	0.00	0.00	0.00	0.00	0.00	0.00	0.00	8.61	0.00	0.00	0.00	0.00	0.00	0.00	0.00
Phyla with <1.5% Representation or No Relative	6.25	1.98	0.58	3.69	1.79	2.98	1.42	1.44	4.81	3.23	2.12	4.31	3.08	0.79	2.11

While some taxa seemed to vary greatly between the two sites, others were very similar. *Chloroflexi*, for example, appeared to have a similar abundance in both Site 1378 and 1379 (Table 1). In addition, the abundance with depth for *Chloroflexi* was also generally stable, ranging from about 13-28% in most samples. One exception was found in the deepest sample examined at Site 1378 (329.1 mbsf) where the proportion was considerably lower than normal at only 5.52% of all sequences. Within that sample, *Elusimicrobia* reached its high at 54.81% of all sequences, but *Spirochaetae* also had a substantial increase, as well as in influx of eukaryal sequences matching to conifers (Tracheophyta; Spermatophyta; Pinophyta; Pinales), likely from either preserved organic remnants or possible contamination. *Chloroflexi* also went higher than typical in one sample, comprising 47.48% of all sequences at 22.08 mbsf, Site 1379. Within that sample, actinobacterial sequences were also found at their highest proportion for Site 1379 (7.85%), and bathyarchaeal sequenced showed a notable drop (18.21%).

Aerophobetes was another phylum that showed similar abundance in Site 1378 as 1379, however the abundance of this phyla with depth was somewhat less stable (Table 1). At Site 1378, its abundance ranged from 0.38% to 20.05%, with the highest amounts occurring at mid-range depths, while at 1379, it ranged from 2.40% to 8.68% from 2.90 to 115.07 mbsf, but reached 36.96% in the deepest 1379 sample at 211.53 mbsf. *Aerophobetes* is one of many taxa found in sub-seafloor sediments for which still little is known about its diversity and metabolic capabilities, though some previous work has provided evidence of acetogenic fermentation in at least some members [38]. *Aerophobetes* sequences could not be classified beyond the phylum level.

Overall, the two sites show a large amount of divergence in their relative compositions. This is surprising, considering that the sites are geographically close (Figure 1), and very similar in lithologies with clay, claystone, silt, siltstone, sand and sandstone, and pore water chemistry profiles, showing sulfate and methane transition zones, with sulfate being removed at 14 mbsf at Site 1378 and 30 mbsf at Site 1379 [5]. As most of the dominant microbial phyla in these sites have very few or no cultured representatives, very little is known about their habitat preferences and possible metabolic strategies [39]. In order to examine possible influences of environmental parameters on community composition at these sites, bivariate correlations were carried out between relative abundance of phyla in the sediment samples and available pore water chemistry variables. Correlations were performed simultaneously for all chemical variables to provide a measure of statistical significance. Significant correlations (<0.05, 2-tailed test) to environmental parameters were found for *Bathyarchaeota, Lokiarchaeota, Planctomycetes, Deltaproteobacteria, Elusimicrobia, Atribacteria,* and *Aerophobetes* (Table S2, Figure 5). Many of the chemistry variables also co-varied with depth and with each other, so depth correlations, to the chemistry variables examined, are also given for reference (Table S2). While some correlations were seen with magnesium, lithium, and boron (Table S2), we excluded them from further analysis due to the fact they are typically uninvolved in anaerobic microbial metabolisms (Figure 5).

For Site 1378, significant correlations were observed between available pore water chemistry data and *Bathyarchaeota, Lokiarchaeaota, Planctomycetes, Deltaproteobacteria, Elusimicrobia* and *Atribacteria* (Table S2; Figure 5). The presence of many significant correlations, which generally all correlated with depth as well, complicates any interpretation. While co-variance with depth does not preclude any significant relationships to microbial phyla, it is difficult to determine from these data alone, whether or not the trend was due to a direct relationship or a consequence of co-variance. The relationship with sulfate, however, was an exception in that sulfate did not correlate significantly with depth, but did show a significant positive correlation with *Bathyarchaeota* (0.984). However, *Bathyarchaeota* are in higher abundance at Site 1379, yet no such relationship with sulfate was seen. Relationships of *Bathyarchaeota* with sulfate or sulfide may only be able to be seen at individual group levels, which are difficult to determine with the short sequences used in this study [40].

The increase in ammonium, with depth in the sediment column, is generally attributed to the breakdown of organic matter by heterotrophic organisms. At Site 1378, as *Bathyarchaeota, Lokiarchaeota, Planctomycetes,* and *Deltaproteobacteria* decline in proportion with depth, they show a negative correlation with ammonium, which is also likely the reason for the *Lokiarchaetoa* and *Deltaproteobacteria* to show negative correlations with bromine, which tends to increase at depth. *Elusimicrobia* and *Atribacteria,* show a positive relationship with ammonium (0.774 and 0.725, respectively), suggesting that they might be a part of the active heterotrophic community at this site (Figure 5). At Site 1379, candidate phylum *Aerophobetes* has a similar positive correlation with ammonia (0.759), and additionally has a concurrent positive relationship with bromine (0.871). The *Aerophobetes* have previously been suggested to be anaerobic autotrophs performing hydrogen oxidation, but more diversity may be found in this group [39]. The correlation in this study could be suggestive of a potential relationship to the degradation of brominated organic matter, a metabolism of interest to the sub-seafloor research community [41–43]. Also at Site 1379, *Planctomycetes* showed a positive correlation with sulfate (0.955), but this is likely due to the decrease in cells with depth, as had been seen for the previous ammonium correlations. Finally at Site 1379, *Lokiarchaeota* and *Bathyarchaeota* correlated with ammonia, in opposite trends, with *Lokiarchaeaota* correlating negatively and *Bathyarchaeota* correlating positively (Table S2; Figure 5).

While significant correlations between pore water chemistry data and microbial phyla are not strong evidence for microbial metabolisms, some of the relationships seen here hint at possible metabolic causes and can guide current efforts underway to rebuild metabolic profiles from the metagenomics datasets in this study. Geologic parameters may also exert influences, including proximity to the subduction zone as Site 1378 is closer, but this is difficult to precisely constrain for this study [5].

A) 1378

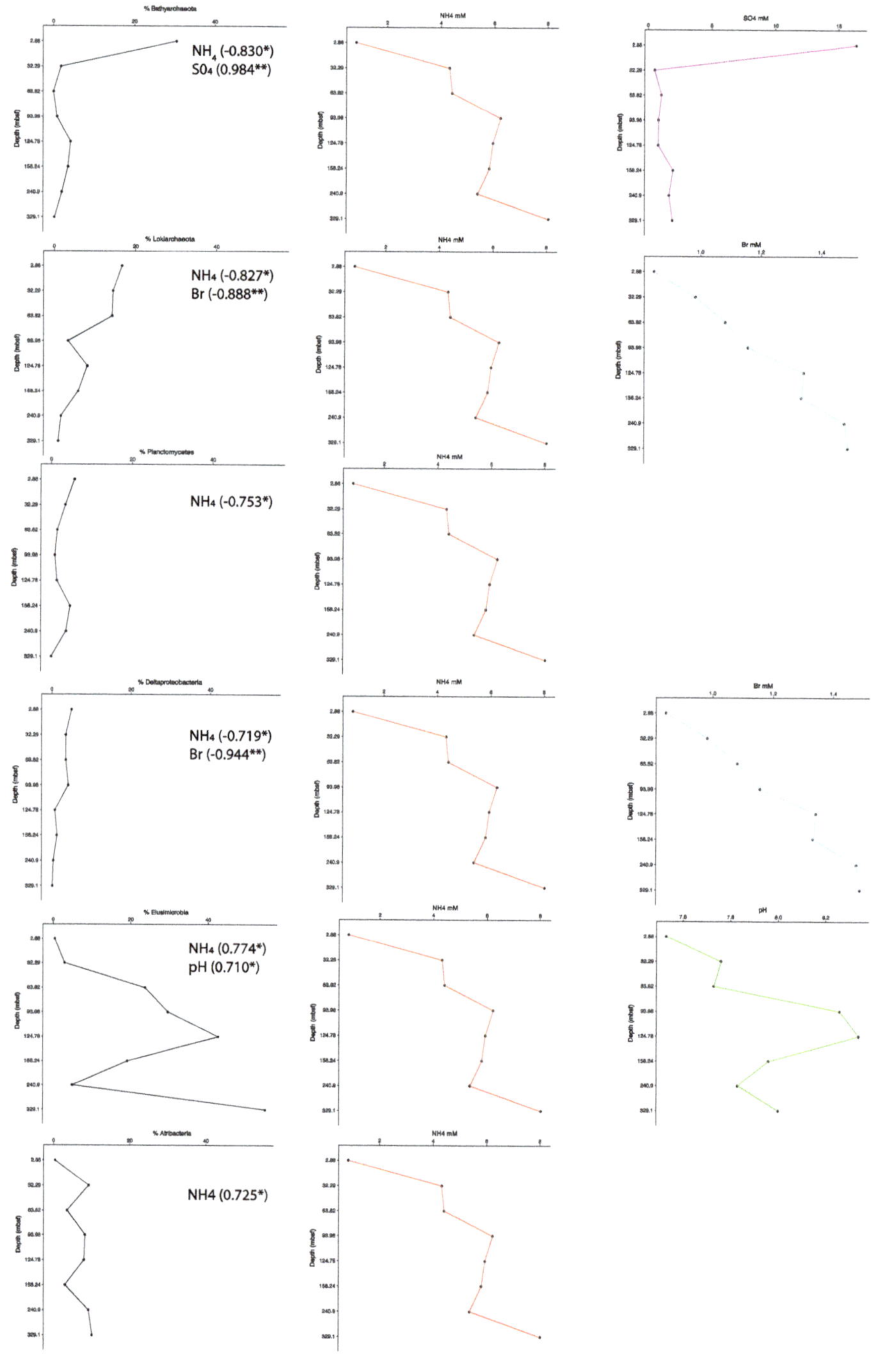

Figure 5. *Cont.*

B) 1379

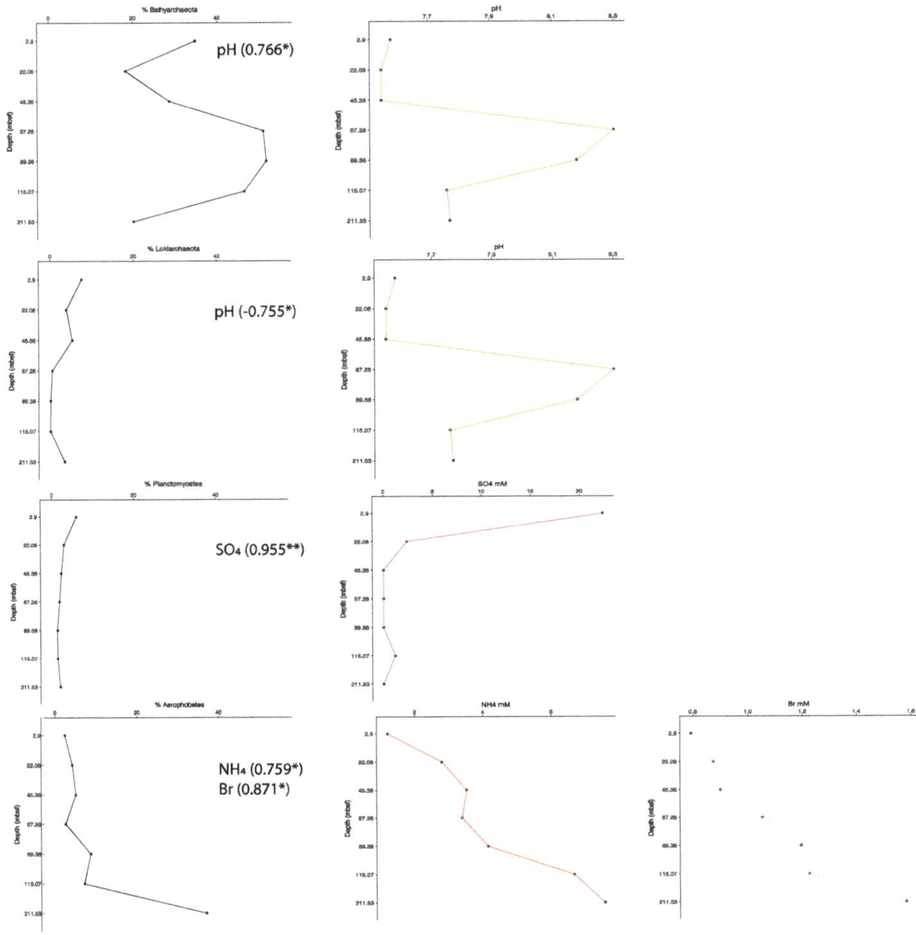

Figure 5. Taxon abundances with depth and their correlated geochemistry for Sites (**A**) 1378 and (**B**) 1379. Significant Pearson correlation values are shown.

3.3. Influence of Drilling Fluid

A sample of drilling fluid was obtained directly from the supply utilized during drilling operations of IODP Expedition 334, which included seawater, as well as a mineral component. The sample was processed and sequenced in the same manner as sediment samples obtained from Sites 1378 and 1379 drilled during this expedition. The classification of sequences from the drilling fluid was examined, and sequences were found to be primarily from *Alphaproteobacteria*, *Gammaproteobacteria*, and *Bacteroidetes* (for full list of classified sequences see Table S3). The sequences in the drilling fluid were then carefully compared with those in the sediment samples in order to assess the potential influence of drilling fluid on microbial community analysis using universal 16S rRNA amplicon sequencing (Figure 6).

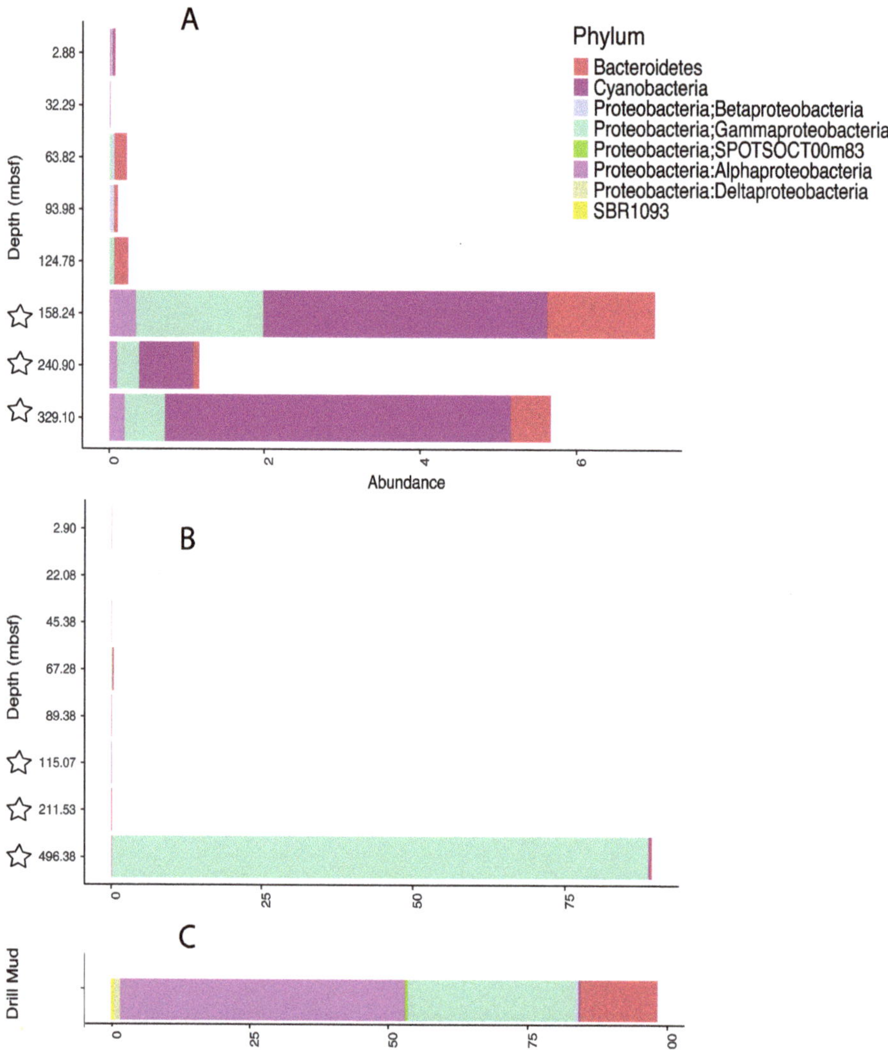

Figure 6. Proportion of classified sequences from each sediment sample that matched those present in the drilling fluid at the taxonomic level of "order." The matched sequences for each order taxon were totaled for each phylum/subphylum depicted above. (**A**) Sequences from Site 1378; (**B**) Sequences from Site 1379; (**C**) Sequences from the drilling mud. Stars represent samples taken from XCB cores. All other were retrieved from APC coring.

For samples taken by the APC coring system, these sequences made up no more than 0.2% of the total classified reads from Site 1378, and no more than 0.4% of total classified reads from Site 1379. At Site 1378, there was a noticeable increase in the proportion of the potential drilling-fluid-sourced sequences in samples taken with XCB coring (158, 241, and 329 mbsf); however, their percentages were still relatively low with respect to the total identified reads (7.0, 1.2, and 5.7 % respectively). At Site 1379, the proportion of potential drilling-fluid-sourced sequences in samples taken with XCB coring remained as low as the APC cored samples from 115 and 212 mbsf (0.5, and 0.10%, respectively), but was very high (89.2%) in the deepest sample (496 mbsf), due to an unusually high number of sequences in one particular lineage of *Gammaproteobacteria*, the genus *Oceanospirillum* (Figure 6).

Significantly, in nearly all samples examined, the contribution of sequences, potentially originating from drilling fluid was relatively small, even in XCB cored samples taken from up to 329 meters below seafloor (where the switchover from APC to XCB occurred at 128 mbsf). There can be exceptions where contamination may coincide with increased drilling difficulty or sediment porosity, potentially seen by the single sample at 496 mbsf that overlapped greatly with potential contaminants (Figure 6). Overall, the results presented here show that, most often, drilling fluid contributes only a minor amount to the community DNA profile, which lends confidence to the conclusions drawn even when it is not possible to remove all ambiguous lineages. Grouping at the level of order, instead of the operational taxonomic unit (OTU), may allow for additional contaminant lineages to be removed even if diversity is not completely determined, without removing many subsurface inhabitants. We suggest that with thorough sampling of contaminating drilling fluid, sediment cores, including the XCB cores of harder sediment packages, can be confidently utilized for biological study with molecular methods.

4. Conclusions

Microbial community analyses of Sites 1378 and 1379 of the Costa Rica margin sub-seafloor show that, at the domain level, both Bacteria and Archaea constitute significant portions of the microbial community—a result confirmed by primer validation efforts and sequencing of metagenomes. Archaea are particularly abundant at Site 1379. Both sites revealed similar proportions of the phyla *Chloroflexi* and *Aerophobetes*; Site 1378 had a notably higher proportion of *Elusimicrobia*, *Actinobacteria*, *Lokiarchaeota* and *Atribacteria*; and Site 1379 had a notably higher proportion of *Bathyarchaeota* and *Hadesarchaeota*. Many of the taxonomic lineages identified in this study have been found often in sub-seafloor environments around the globe, but changes in their relative proportions, depending on their specific location, suggest that they may be influenced by variations not yet observed in environmental parameters within the sub-seafloor. Correlations to pore water chemistry, included here, revealed some possible metabolic strategies for sub-seafloor lineages, including the importance of heterotrophic activity on ammonia concentrations, and these relationships are being explored via the continued analysis of metagenomic datasets presented in this study. Finally, we suggest that drilling contamination, while an issue that needs routine monitoring, is not a major hindrance for the molecular analysis of microbial community composition in deep sub-seafloor samples.

Supplementary Materials: The following are available online at http://www.mdpi.com/2076-3263/9/5/218/s1, Supporting methods; Table S1: Sample information; Table S2: Correlation values; Table S3: Drilling fluid composition.

Author Contributions: Conceptualization, A.M. and C.H.H.; methodology, A.M, M.E.R., and R.L.-Z.; data curation, A.M, M.E.R., R.L.-Z., and I.E.-V.; writing—original draft preparation, A.M., R.L.-Z., and J.F.B.; writing—review and editing, A.M. and J.F.B.; funding acquisition, J.F.B. and C.H.H.

Funding: This work was funded by IODP 334 Post Expedition award T334A40, the Penn State Astrobiology Research Center (through the NASA Astrobiology Institute, cooperative agreement #NNA09DA76A), the Pennsylvania Space Grant Consortium (NNX10AK74H), and a postdoctoral fellowship (RLZ) and graduate student fellowship (AJM) from NSF-funded Center for Dark Energy Biosphere investigations.

Acknowledgments: We thank all shipboard scientists on Expedition 334, especially geochemist Evan Solomon for providing refined pore water chemistry data. Illumina sequencing data for Site 1378 were made possible by the Deep Carbon Observatory's Census of Deep Life, supported by the Alfred P. Sloan Foundation. This Illumina sequencing was performed at the Marine Biological Laboratory (Woods Hole, MA, USA) and we are grateful for the assistance of Mitch Sogin, Susan Huse, Joseph Vineis, Andrew Voorhis, Sharon Grim, and Hilary Morrison at the Marine Biological Laboratory. Illumina sequencing for Site 1379 was carried out by the lab of Stephan Schuster, and we are thankful for the assistance of Lynn Tomsho. Additionally, we thank Glenn Christman for metagenome assembly and bioinformatics assistance. This work is C-DEBI contribution 473.

References

1. D'Hondt, S.; Jorgensen, B.B.; Miller, D.J.; Batzke, A.; Blake, R.; Cragg, B.A.; Cypionka, H.; Dickens, G.R.; Ferdelan, T.; Hinrichs, K.U.; et al. Distributions of microbial activities in deep subseafloor sediments. *Science* **2004**, *306*, 2216–2221. [CrossRef] [PubMed]

2. Ciobanu, M.C.; Burgaud, G.; Dufresne, A.; Breuker, A.; Redou, V.; Maamar, S.B.; Gaboyer, F.; Vandenabeele-Trambouze, O.; Lipp, J.S.; Schippers, A.; et al. Microorganims persist at record depths in the subseafloor of the Canterbury Basin. *ISMEJ* **2014**, *8*, 1370–1380. [CrossRef] [PubMed]

3. Inagaki, F.; Nunoura, T.; Nakagawa, S.; Teske, A.; Lever, M.; Lauer, A.; Suzuki, M.; Takai, K.; Delwiche, M.; Colwell, F.S.; et al. Biogeographical distribution and diversity of microbes in methane hydrate-bearing deep marine sediments on the Pacific Ocean Margin. *Proc. Natl. Acad. Sci. USA* **2006**, *103*, 2815–2820. [CrossRef]

4. Bidle, K.A.; Kastner, M.; Bartlett, D.H. A phylogenetic analysis of microbial communities associated with methane hydrate containing marine fluids and sediments in the Cascadia Margin (ODP site 892B). *FEMS Microbiol. Lett.* **1999**, *177*, 101–108.

5. Vannucchi, P.; Ujiie, K.; Stroncik, N.; the Expedition 334 Scientists. Proceedings of the Integrated Ocean Drilling Program Volume 334. Available online: http://publications.iodp.org/proceedings/334/334title.htm (accessed on 31 March 2019).

6. Barry, P.H.; de Moor, J.M.; Giovanelli, D.; Schrenk, M.; Hummer, D.R.; Lopez, T.; Pratt, C.A.; Alpizar Segura, Y.; Battaglia, A.; Beaudry, P.; et al. Forearc carbon sink reduces long-term volatile recycling into the mantle. *Nature* **2019**, *568*, 487–492. [CrossRef]

7. Schippers, A.; Neretin, L.N.; Kallmeyer, J.; Ferdelman, T.G.; Cragg, B.A.; Parkes, R.J.; Jørgensen, B.B. Prokaryotic cells of the deep sub-seafloor biosphere identified as living bacteria. *Nature* **2005**, *433*, 861–864. [CrossRef] [PubMed]

8. Briggs, B.R.; Inagaki, F.; Morono, Y.; Futagami, T.; Huguet, C.; Rosell-Mele, A.; Lorenson, T.D.; Colwell, F.S. Bacterial dominance in subseafloor sediments characterized by methane hydrates. *FEMS Microbiol. Ecol.* **2012**, *81*, 88–98. [CrossRef] [PubMed]

9. Biddle, J.F.; Fitz-Gibbon, S.; Schuster, S.C.; Brenchley, J.E.; House, C.H. Metagenomic signatures of the Peru Margin subseafloor biosphere show a genetically distinct environment. *Proc. Natl. Acad. Sci. USA* **2008**, *105*, 10583–10588. [CrossRef] [PubMed]

10. Lipp, J.S.; Morono, Y.; Inagaki, F.; Hinrichs, K.-U. Significant contribution of Archaea to extant biomass in marine subsurface sediments. *Nature* **2008**, *454*, 991–994.

11. Smith, D.C.; Spivack, A.J.; Fisk, M.R.; Haveman, S.A.; Staudigel, H.; ODP Leg 185 Shipboard Scientific Party. Methods for Quantifying Potential Microbial Contamination During Deep Ocean Coring. 2000. Available online: http://www-odp.tamu.edu/publications/tnotes/tn28/INDEX.HTM (accessed on 31 March 2019).

12. Smith, D.C.; Spivack, A.J.; Fisk, M.R.; Haveman, S.A.; Staudigel, H.; ODP Leg 185 Shipboard Scientific Party. Tracer-Based Estimates of Drilling-Induced Microbial Contamination of Deep Sea Crust. *Geomicrobiol. J.* **2000**, *17*, 207–219.

13. House, C.H.; Cragg, B.A.; Teske, A. Drilling contamination tests on ODP Leg 201 using chemical and particulate tracers. In Proceedings of the Ocean Drilling Program; Texas A&M University: College Station, TX, USA, 2003; pp. 1–19.

14. Masui, N.; Morono, Y.; Inagaki, F. Microbiological assessment of circulation mud fluids during the first operation of riser drilling by the deep-earth research vessel Chikyu. *Geomicrobiol. J.* **2008**, *25*, 274–282. [CrossRef]

15. Santelli, C.M.; Banerjee, N.; Bach, W.; Edwards, K.J. Tapping the subsurface ocean crust biosphere: Low biomass and drilling-related contamination calls for improved quality controls. *Geomicrobiol. J.* **2010**, *27*, 158–169. [CrossRef]

16. Fry, J.C.; Parkes, R.J.; Cragg, B.C.; Weightman, A.J.; Webster, G. Prokaryotic biodiversity and activity in the deep subseafloor biosphere. *FEMS Microbiol. Ecol.* **2008**, *66*, 181–196. [CrossRef] [PubMed]

17. Orcutt, B.N.; Sylvan, J.B.; Knab, N.J.; Edwards, K.J. Microbial ecology of the dark ocean above, at, and below the seafloor. *Microbiol. Mol. Biol. Rev.* **2011**, *75*, 361–422. [CrossRef]

18. Inagaki, F.; Hinrichs, K.-U.; Kubo, Y.; Bowles, M.W.; Heuer, V.B.; Hong, W.-L.; Hoshino, T.; Ijiri, A.; Imachi, H.; Ito, M.; et al. Exploring deep microbial life in coal-bearing sediment down to ~2.5 km below the ocean floor. *Science* **2015**, *349*, 420–424. [CrossRef] [PubMed]

19. Yanagawa, K.; Nunoura, T.; McAllister, S.M.; Kirai, M.; Breuker, A.; Brandt, L.; House, C.H.; Moyer, C.L.; Birrien, J.L.; Aoike, K.; et al. The first microbiological contamination assessment by deep-sea drilling and coring by the D/V Chikyu at the Iheya North hydrothermal field in the Mid-Okinawa Trough (IODP Expedition 331). *Frontiers Microbiol.* **2013**, *4*, 327. [CrossRef] [PubMed]

20. Stavenhagen, A.U.; Flueh, E.R.; Ranero, C.R.; McIntosh, K.D.; Shipley, T.; Leandro, G.; Schulze, A.; Dañobeitia, J.J. Seismic wide-angle investigations in Costa Rica—A crustal velocity model from the Pacific to the Caribbean. ZBL. *Geol. Paläontol.* **1998**, *1*, 393–408.

21. Rhodes, M.E.; Fitz-Gibbon, S.T.; Oren, A.; House, C.H. Amino acid signatures of salinity on an environmental scale with a focus on the Dead Sea. *Environ. Microbiol.* **2010**, *12*, 2613–2623. [CrossRef] [PubMed]

22. Schloss, P.D.; Westcott, S.L.; Ryabin, T.; Hall, J.R.; Hartmann, M.; Hollister, E.B.; Lesniewski, R.A.; Oakley, B.B.; Parks, D.H.; Robinson, C.J.; et al. Introducing mothur: Open-source, platform-independent, community-supported software for describing and comparing microbial communities. *Appl. Environ. Microbiol.* **2009**, *75*, 7537–7541. [CrossRef]

23. Quast, C.; Pruesse, E.; Yilmaz, P.; Gerken, J.; Schweer, T.; Yarza, P.; Peplies, J.; Glockner, F.O. The SILVA ribosomal RNA gene database project: Improved data processing and web-based tools. *Nucleic Acids Res.* **2013**, *41*, 590–596. [CrossRef]

24. Pruesse, E.; Peplies, J.; Glöckner, F.O. SINA: Accurate high-throughput multiple sequence alignment of ribosomal RNA genes. *Bioinformatics* **2012**, *28*, 1823–1829. [CrossRef]

25. Li, W.; Godzik, A. Cd-hit: A fast program for clustering and comparing large sets of protein or nucleotide sequences. *Bioinformatics* **2006**, *22*, 1658–1659. [CrossRef]

26. Silva. Available online: http://www.arb-silva.de (accessed on 31 March 2019).

27. BLAST: Basic Local Alignment Search Tool. Available online: http://blast.ncbi.blm.nih.gov/Blast.cgi (accessed on 31 March 2019).

28. Camacho, C.; Coulouris, G.; Avagyan, V.; Ma, N.; Papadopoulos, J.; Bealer, K.; Madden, T. BLAST+: Architecture and applications. *BMC Bioinform.* **2009**, *10*, 421. [CrossRef]

29. Schmieder, R.; Edwards, R. Quality control and preprocessing of metagenomic datasets. *Bioinformatics* **2011**, *27*, 863–864. [CrossRef]

30. Bolger, A.M.; Lohse, M.; Usadel, B. Trimmomatic: A flexible trimmer for Illumina sequence data. *Bioinformatics* **2014**, *30*, 2114–2120. [CrossRef]

31. Menzel, P.; Ng, K.L.; Krogh, A. Fast and sensitive taxonomic classification for metagenomics with Kaiju. *Nature Commun.* **2016**, *7*, 11257. [CrossRef] [PubMed]

32. Darling, A.E.; Jospin, G.; Lowe, E.; Matsen, F.A.; Bik, H.M.; Eisen, J.A. PhyloSift: Phylogenetic analysis of genomes and metagenomes. *PeerJ* **2014**, *9*, 2:e243. [CrossRef]

33. Berg, R.; Solomon, E.A. Geochemical constraints on the distribution and rates of debromination in the deep subseafloor biosphere. *Geochim. Cosmochim. Acta* **2016**, *174*, 30–41. [CrossRef]

34. Torres, M.E.; Muratli, J.M.; Solomon, E.A. Data Report: Minor Element Concentrations in Pore Fluids from the CRISP-A Transect Drilled During Expedition 334. Available online: http://publications.iodp.org/proceedings/334/201/201_.htm (accessed on 31 March 2019).

35. Rhodes, M.E.; Oren, A.; House, C.H. Dynamics and persistence of Dead Sea microbial populations as shown by high-throughput sequencing of rRNA. *Appl. Environ. Microbiol.* **2012**, *78*, 2489–2492. [CrossRef]

36. Zhou, Z.; Pan, J.; Wang, F.; Gu, J.D.; Li, M. Bathyarchaeota: Globally distributed metabolic generalists in anoxic environments. *FEMS Microbiol. Lett.* **2018**, *42*, 639–655. [CrossRef]

37. Geissinger, O.; Herlemann, D.P.R.; Morschel, E.; Maier, U.G.; Brune, A. The ultramicrobacterium "Elusimicrobium minutum" gen. nov., sp. nov., the first cultivated representative of the termite group 1 phylum. *Appl. Env. Microbiol.* **2009**, *75*, 2831–2840. [CrossRef]

38. Wang, Q.; Garrity, G.M.; Tiedje, J.M.; Cole, J.R. Native Bayesian classifier for rapid assignment of rRNA sequences into the new bacterial taxonomy. *Appl. Environ. Microbiol.* **2007**, *73*, 5261–5267. [CrossRef] [PubMed]

39. Rinke, C.; Schwientek, P.; Sczyrba, A.; Ivanova, N.N.; Anderson, I.J.; Cheng, J.-F.; Darling, A.; Malfatti, S.; Swan, B.K.; Gies, E.A.; et al. Insights into the phylogeny and coding potential of microbial dark matter. *Nature* **2013**, *499*, 431–437. [CrossRef]

40. Lazar, C.S.; Biddle, J.F.; Meador, T.B.; Blair, N.; Hinrichs, K.-U.; Teske, A.P. Environmental controls on intragroup diversity of the uncultured benthic archaea of the miscellaneous Crenarchaeotal group lineage naturally enriched in anoxic sediments of the White Oak River estuary (North Caroline, USA). *Env. Microbiol.* **2015**, *17*, 2228–2238. [CrossRef] [PubMed]

41. Futagami, T.; Morono, Y.; Terada, T.; Kaksonen, A.H.; Inagaki, F. Dehalogenation activities and distribution of reductive dehalogenase homologous genes in marine subsurface sediments. *Appl. Environ. Microbiol.* **2009**, *75*, 6905–6909. [CrossRef] [PubMed]

42. Futagami, T.; Morono, Y.; Terada, T.; Kaksonen, A.H.; Inagaki, F. Distribution of dehalogenation activity in subseafloor sediments of the Nankai Trough subduction zone. *Philos. Trans. R Soc. Lond. B Biol. Sci.* **2013**, *368*, 20120249. [CrossRef] [PubMed]

43. Kawai, M.; Futagami, T.; Toyoda, A.; Takaki, Y.; Nishi, S.; Hori, S.; Arai, W.; Tsubouchi, T.; Morono, Y.; Uchiyama, I.; et al. High frequency of phylogenetically diverse reductive dehalogenase-homologous genes in deep subseafloor sedimentary metagenomes. *Front. Microbiol.* **2014**, 5. [CrossRef] [PubMed]

Article

Re-Evaluating the Age of Deep Biosphere Fossils in the Lockne Impact Structure

Mikael Tillberg [1,2,*], Magnus Ivarsson [3,4], Henrik Drake [1], Martin J. Whitehouse [5], Ellen Kooijman [5] and Melanie Schmitt [5]

[1]   Department of Biology and Environmental Science, Linnaeus University, 392 31 Kalmar, Sweden; Henrik.drake@lnu.se
[2]   Department of Earth Sciences, University of Gothenburg, Box 460, 40530 Göteborg, Sweden
[3]   Department of Biology, University of Southern Denmark, Campusvej 55, 5230 Odense, Denmark; Magnus.ivarsson@nrm.se
[4]   Department of Paleobiology, Swedish Museum of Natural History, Box 50 007, 104 05 Stockholm, Sweden
[5]   Department of Geosciences, Swedish Museum of Natural History, Box 50 007, 104 05 Stockholm, Sweden; martin.whitehouse@nrm.se (M.J.W.); ellen.kooijman@nrm.se (E.K.); Melanie.schmitt@nrm.se (M.S.)
*   Correspondence: mikael.tillberg@lnu.se

Received: 15 March 2019; Accepted: 4 May 2019; Published: 7 May 2019

Abstract: Impact-generated hydrothermal systems have been suggested as favourable environments for deep microbial ecosystems on Earth, and possibly beyond. Fossil evidence from a handful of impact craters worldwide have been used to support this notion. However, as always with mineralized remains of microorganisms in crystalline rock, certain time constraints with respect to the ecosystems and their subsequent fossilization are difficult to obtain. Here we re-evaluate previously described fungal fossils from the Lockne crater (458 Ma), Sweden. Based on in-situ Rb/Sr dating of secondary calcite-albite-feldspar (356.6 ± 6.7 Ma) we conclude that the fungal colonization took place at least 100 Myr after the impact event, thus long after the impact-induced hydrothermal activity ceased. We also present microscale stable isotope data of ^{13}C-enriched calcite suggesting the presence of methanogens contemporary with the fungi. Thus, the Lockne fungi fossils are not, as previously thought, related to the impact event, but nevertheless have colonized fractures that may have been formed or were reactivated by the impact. Instead, the Lockne fossils show similar features as recent findings of ancient microbial remains elsewhere in the fractured Swedish Precambrian basement and may thus represent a more general feature in this scarcely explored habitat than previously known.

Keywords: Impact structure; fungal hyphae; *in situ* radiometric dating; secondary minerals; stable isotopes

---

1. Introduction

Impact craters and associated impact-generated hydrothermal systems have been suggested as favourable environments for microbial life [1–4]. Extensive fracturing at depth caused by impacts provides pore space for endolithic communities, and heat generated by the impact drives hydrothermal convection favourable for deep ecosystems [5,6]. Depending on the size and extension of the impact melt, heat can sustain for hundreds of thousands, up to millions of years [7–11]. A handful of reports of fossil- and geochemical signatures support post-impact colonization of impact paleo-hydrothermal systems. Sulfur isotope signatures of sulfates from the Haughton crater, Canada, suggest rapid colonization of the impact-generated hydrothermal system by sulfate-reducing bacteria [12]. Tubular ichnofossils in impact glass from the Ries crater, Germany, indicate microbial activity in a post-impact hydrothermal system [13]. The presence of putative fossilized microorganisms in samples from the Dellen crater in Sweden [14] as well as fossilized biofilm in the Siljan crater [15], suggests colonization of the hydrothermal systems, respectively. Exceptionally well preserved fungal fossils were described in drill cores at a depth of ~170 to 220 meters below the surface from the Lockne crater (458 Ma), Sweden [16]. The fungal fossils were preserved as carbonaceous matter, and displayed characteristic fungal morphologies such as repetitive septa, frequent branching and anastomosis between branches resulting in mycelium-like communities. The close association with oils and minerals interpreted as products of the hydrothermal activity was used to relate the fossils to the post-impact hydrothermal system, presumed to have prevailed about 10,000 years after the impact, and thus indirectly providing an age constraint for the fossils.

With the exception of the subseafloor igneous crust, the fossil archive of the deep biosphere is generally less explored as its living and modern counterpart [17] and age determination of those fossils is often associated with large uncertainties. Commonly, ages of fossils are associated with the radiometric age of the host rock and/or to major tectonic events affecting the rock, which, however, cannot directly be assigned to microbial colonization and subsequent fossilization [18]. In recent years, application of in situ radiometric dating of authigenic minerals produced by the ancient microbial community have been demonstrated to be successful in gaining time constraints on the colonization, both by utilization of U-Pb dating of secondary carbonate [19] and Rb/Sr dating of secondary calcite/fluorite-feldspar and calcite-clay minerals [19–22]. In situ stable carbon isotope analysis of secondary calcite (δ^{13}C), sulfide and sulfate (δ^{34}S, mainly for pyrite, but also chalcopyrite and barite) have been used to understand the microbial processes (e.g., methane oxidation, methanogenesis and microbial sulfate reduction) in deep fracture systems within the continental crystalline crust [23–26].

Here, we present in situ dating of secondary vein minerals associated with the Lockne fungal fossils [10], which challenges the fossil association with the impact-generated system, and suggests that fracture-reactivation due to hydrothermal activity occurred in the fracture system long after the impact (100 Ma) and thus infers a previously unknown, younger maximum age, for the fossils. In addition, we present microscale stable isotope data of ^{13}C-enriched calcite suggesting the presence of methanogens contemporary with the fungi. Collectively, the Lockne community bears close similarities to other fossil fungal-prokaryotic consortia described in Swedish crystalline basement [23,27].

2. Geological Setting

The Lockne impact crater, Sweden, is a concentric structure with an inner crater diameter of 7.5 km developed in the crystalline basement. The oldest post-impact sediments have been dated at 458 Ma [28]. At the time of impact, the area was covered by a more than 500 m deep sea [29]. The targets were the 1.86–1.85 Ga Revsund granitoids [30] and the 1250–1200 Ma alkali- and olivine-rich Åsby dolerite [31] overlain by a sedimentary cover of Cambrian bituminous black mud (today alum shale) and Ordovician consolidated limestone [32].

The drill core used in this study intersects the north-western part of the Lockne impact structure [33]. Structurally, this represents the marginal, shallower rim of the final crater, and not the deeper central parts. The cored sequence comprises crystalline impact breccia and fractured basement rock, slump- and resurge deposits, and secular sediments. Veins and vugs in the brecciated basement rock are partly filled with hydrothermally formed calcite and quartz and to minor extent with pyrite, chalcopyrite, galena and sphalerite [33]. Bitumen is abundant and associated with kerogenous matter that covers the hydrothermal minerals as a thin film. This film is a couple of micrometers thick and has a yellow-brownish appearance in optical microscopy. The C-rich film is associated with long, undulating and curvilinear filaments that are preserved as complex networks forming entangled, almost chaotic assemblages. Based on morphological traits identical to fungi the filaments were interpreted as fossilized hyphae forming mycelia in the open pore space of the granitoids [16]. The fungi were introduced into the system after the impact, and colonized the secondary mineralizations by the formation of an initial biofilm from which hyphae grew and formed the mycelium (Figure 1). The fungi were described to have colonized the rock during the hydrothermal activity and upon death become impregnated by oils migrating through the hydrothermal system. This process eventually preserved the fungal mycelia as carbonaceous fossils [16].

3. Sample Material

Mineral samples with fungal mycelia (Figure 1) were taken from drill core LOC1 (total length 225.15 m), and are the samples as presented in [16]. Bitumen occurs in a section between 171.30 and 219.90 m depth in the core in veins and vugs in the fractured and brecciated basement rock. A carbonaceous film from which fungal hyphae propagate and form complex mycelium-like networks covers many euhedral blocky (short c-axis) calcite crystals that occur together with quartz, K-feldspar (adularia), albite and chalcopyrite.

4. Methods

Crystals were physically removed from the fracture voids of the impact breccia, and individually embedded in epoxy. The epoxy mounts were polished and examined by a scanning electron microscope (SEM, Hitachi) equipped with an energy-dispersive spectrometer (EDS), at the University of Gothenburg, Sweden. Sample characterization of the polished crystals was carried out before isotope analysis described below.

4.1. Secondary Ion Mass Spectrometry (SIMS) Stable Isotope Analysis

Calcite and chalcopyrite crystals were mounted in epoxy, polished to expose cross-sections and examined with SEM to trace zonations and impurities prior to SIMS analysis. Intra-crystal SIMS-analysis (10 μm lateral beam dimension, 1-2 μm depth dimension) of sulfur isotopes in chalcopyrite and carbon isotopes in calcite was performed on a CAMECA IMS1280 ion microprobe (CAMECA, SAS, Gennevilliers, France) following the analytical settings and tuning reported previously [19,20]. Sulfur was sputtered using a $^{133}Cs^+$ primary beam with 20 kV incident energy (10 kV primary, −10 kV secondary) and a primary beam current of ~1.5 nA. A normal incidence electron gun was used for charge compensation. Analyses were performed in automated sequences, with each analysis comprising a 70 second pre-sputter to remove the gold coating over a rastered 15 × 15 μm area, centering of the secondary beam in the field aperture to correct for small variations in surface relief and data acquisition in sixteen four second integration cycles. The magnetic field was locked at the beginning of the session using a nuclear magnetic resonance (NMR) field sensor. Secondary ion signals for ^{32}S and ^{34}S were detected simultaneously using two Faraday detectors with a common mass resolution of 4860 (M/ΔM). Data were normalised for instrumental mass fractionation using matrix-matched reference materials which were mounted together with the sample mounts and analysed after every sixth sample analysis. Results are reported as per mil (‰) $\delta^{34}S$ based on the Canon Diablo Troilite (V-CDT)-reference value. Up to 17 crystals were analysed from each fracture sample. In total, 37 analyses were made for $\delta^{34}S$ ($^{34}S/^{32}S$) of chalcopyrite from 15 crystals from fracture-coating sample 185 m. The Trout Lake chalcopyrite reference material with a conventionally determined value of +0.3‰ [34] was used. Typical precision on a single $\delta^{34}S$ value, after propagating the within run and external uncertainties from the reference material measurements was ±0.08‰.

For calcite, 13 $\delta^{13}C$ SIMS-analyses were performed on the same CAMECA IMS1280 described above. Settings follow those described for S isotopes, with some differences: a Faraday cage/electron multiplier (FC/EM) combination with mass resolution 2500 on the ^{12}C peak and 4000 on the ^{13}C peak was used to resolve it from $^{12}C^1H$. Influence of organic carbon was avoided by careful spot placement to areas in the crystals without micro-fractures or inclusions and at a sufficient distance from grain-boundaries where fine-grained clusters of other minerals and remnants of organic material may appear. The uncertainty associated with potential organic inclusions and matrix composition is therefore considered to be insignificant compared to the isotopic variations. Calcite results are reported as per mil (‰) $\delta^{13}C$ based on the Pee Dee Belemnite (V-PDB) reference value. Analyses were carried out running blocks of six unknowns bracketed by two reference material analyses. Analytical transects of up to six spots were made from core to rim in the crystals. Analyses were made in three crystals from two fracture samples (181 and 185 m). Isotope data from calcite were normalised using calcite reference material S0161 from a granulite facies marble in the Adirondack Mountains, kindly provided by R.A. Stern (University of Alberta). The values used for IMF correction were determined by conventional stable isotope mass spectrometry at Stockholm University on ten separate pieces, yielding $\delta^{13}C = 0.22 \pm 0.11$‰ V-PDB (1 std. dev.). Precision was $\delta^{13}C$: ± 0.4–0.5‰. Values of the reference material measurements are listed together with the samples in Tables 1 and 2.

4.2. Laser Ablation Multi-Collector Inductively Coupled Plasma Mass Spectrometry (LA-MC-ICP-MS) $^{87}Sr/^{86}Sr$

The $^{87}Sr/^{86}Sr$ values of calcite in sample 185 m were determined by laser ablation multi-collector inductively coupled plasma mass spectrometry (LA-MC-ICP-MS) analysis at the Vegacenter, Swedish Museum of Natural History, Stockholm, Sweden, using a Nu plasma II MC-ICP-MS (Nu Instruments Ltd, Wrexham, UK), and an electrospray ionization (ESI) NWR193 ArF excimer laser ablation system (Elemental Scientific Lasers, Bozeman, MT, USA). Four of the six spots analysed for $\delta^{13}C$ using SIMS were also analysed for Sr isotopes by LA-ICP-MS (using larger spots). Ablation frequency was 15 Hz, spot size was 80 µm and fluence was 2.8 J/cm^2. Samples were ablated for 45 seconds, followed by 45 seconds wash-out time. The $^{87}Sr/^{86}Sr$ analyses were normalised to an in-house brachiopod reference material 'Ecnomiosa gerda' (linear drift and accuracy correction) using a value established by thermal ionisation mass spectrometry (TIMS) of 0.709181 (2sd 0.000004, [35]). A modern oyster shell from Western Australia was used as a secondary reference material and analysed at regular intervals together with the primary reference. The accuracy of these analyses was quantified by comparison to the modern seawater value for $^{87}Sr/^{86}Sr$ of 0.7091792 $\pm$ 0.0000021 [36]. Values of the reference material measurements are listed together with the sample data in Table 2.

4.3. LA-ICP-MS Analyses for Rb-Sr Dating

The Rb-Sr dating system builds on the beta-decay of ^{87}Rb to ^{87}Sr in minerals. One or several Rb-rich minerals (showing increased $^{87}Sr/^{86}Sr$ and decreased $^{87}Rb/^{86}Sr$ with time) along with a co-genetic Sr-rich mineral (constant $^{87}Sr/^{86}Sr$ with time), in our case secondary adularia in paragenesis with calcite or albite, were analysed by Rb-Sr geochronology via high spatial resolution LA-ICP-MS [37] at the Earth Sciences Centre, University of Gothenburg, Sweden. The Rb/Sr spot analyses in fine-grained adularia (n = 10), calcite (n = 4) and albite (n = 6 (of which 3 = rejected)) from sample 185 m and adularia from sample 216 m (n = 13) were performed using an ESI 213NWR laser ablation system (Elemental Scientific Lasers, Bozeman, MT, USA) connected to Agilent 8800QQQ ICP-MS (Agilent Technologies Inc, Santa Clara, CA, USA) with an ORS3 octopole reaction system reaction cell sandwiched between two quadrupoles. Following laser warm-up, ablation occurred with static spot mode in a constant He flow (800 mL/min). The ablated material was mixed with N_2 and Ar before entering the ICP-MS torch region and reacted with N_2O gas in the reaction cell to chemically separate ^{87}Rb from ^{87}Sr by producing oxide of ^{87}Sr, and thereby enable calculation of $^{87}Rb/^{86}Sr$ and $^{87}Sr/^{86}Sr$ ratios [38]. While the octopole bias was set to negative voltage, N_2O flow rates in the reaction cell were varied to obtain optimal SrO$^+$ production rates. In tandem mass spectrometry (MS/MS) mode both quadrupoles were controlled while reactive gas was in the reaction cell with the quadrupoles set at different masses to measure reaction products in mass-shifted mode. On mass ^{85}Rb/mass-shifted ^{86}Sr and mass-shifted ^{87}Sr/mass-shifted ^{86}Sr raw ratios were used to calculate $^{87}Rb/^{86}Sr$ and $^{87}Sr/^{86}Sr$, respectively. ^{85}Rb is used as a proxy for ^{87}Rb as it is constant on Earth and within 0.02–0.05% [39]. The raw ratios were converted by correction factors derived from repeated analysis of reference materials NIST SRM 610 and BCR-2G, which are documented to be feasible for calibration of in situ LA-ICP-MS/MS Rb-Sr isotopic data [37,38]. The reference materials were selected to ensure that the pulse/analog setting of each measured isotope was identical in samples and reference materials. The NIST SRM 610 certified reference material with $^{87}Sr/^{86}Sr$ of 0.7097 [40] was used for $^{87}Sr/^{86}Sr$ ratio calibration of the sample data.

^{87}Rb/^{86}Sr calibration was performed by using glass reference material BCR-2G [41], shown in Table 3. For ^{87}Rb/^{86}Sr and ^{87}Sr/^{86}Sr respectively, BCR-2G yielded precisions at 1.12% and 0.21%, while NIST SRM 610 yielded 1.16% and 0.35%. The secondary reference material was LP01, a sample constituting mm-sized euhedral biotite from granodiorite of the La Posta intrusion, California. As suggested by [37], we use a weighted mean age of 91.6 ± 1.2 Ma derived from U-Pb TIMS, Ar-Ar and Rb-Sr studies. The isochron age of LP01 in the analytical session was 94.2 ± 2.9 Ma (n = 13, Table 3). The resulting ages are isochron model fits constructed using the Rb decay constant of Villa et al. [42]. Rho (ρ) values for each spot were calculated using ^{85}Rb/^{86}Sr, ^{87}Sr/^{86}Sr and ^{85}Rb/^{87}Sr ratios (Table 4). Average count rate calculation of reference material data is conducted by Glitter$^©$, whereas sample data reduction and within-run error calculation of important element and isotopic ratios is performed using an in-house spreadsheet. Isotopic homogeneity is ensured through evaluation of analytically reliable laser ablation signals on the scales of single spots, grains and assemblages (justification of rejected spots given in Table 4). No error propagation from uncertainties in literature data or within-run errors of reference materials is applied to sample errors as internally calculated errors are significantly larger as the established 1.5% external errors of this method [38].

5. Results

Polished cross sections reveal growth zonation in the calcite crystals (Figure 2c). Ivarsson et al. [16] interpreted the fossils propagating from calcite crystals as fungi based on the size and appearance of the filaments and mycelia, as well as the presence of fungal characteristic morphologies like frequent branching, anastomoses between branches and repeated septa. All fungal fossils including the film are preserved as disordered carbonaceous matter according to Raman spectroscopy and Time-of-Flight-SIMS analyses [16].

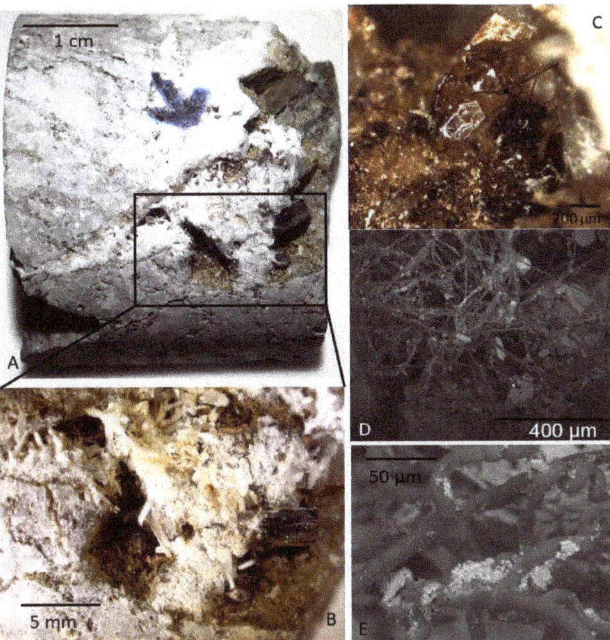

Figure 1. (**A**) Photograph of a piece of drill core from 181.7 meters depth. (**B**) Close up from (A) of a vug containing fossilized mycelia. (**C**) Microphotograph of a calcite covered by a carbonaceous biofilm from which black hyphae is protruding (black arrow). (**D**) Environmental scanning electron microscope (ESEM) image of fungal mycelia. (**E**) ESEM image of branching and coiled hyphae.

5.1. Isotope Compositions

The overall variability of the $\delta^{13}C$ in calcite is large, (Table 2, Figure 2), ranging from ^{13}C-depleted ($-19.4‰$, V-PDB) to ^{13}C-enriched values ($+12.3‰$, V-PDB). The $\delta^{13}C$ composition is, however, highly sample specific, and even crystal specific. Sample 181 m has ^{13}C-depleted values: Crystal 1 has values of $-8.8‰$ to $-6.7‰$, and crystal 2 $-19.4‰$ to $-16.9‰$. In contrast, the crystal analysed in sample 185 m has ^{13}C-enriched composition throughout all spots (n = 6): $+9.9‰$ to $+12.7‰$, as shown in the spot transects (Figure 2d). The correlation of the isotopic variance to the growth zonation is related to minor temporal chemical and isotopic fluctuation in the precipitating fluids. The $^{87}Sr/^{86}Sr$ values of the calcite crystal in sample 185 m also shows small differences of 0.7444 ± 0.0015 to 0.7471 ± 0.0012 (Figure 2d). The $\delta^{34}S_{chalcopyrite}$ values in sample 185 m, show a span from $+0.4‰$ to $+3.6‰$ V-CDT (n = 37, Table 1), and there is no significant variation within the crystals targeted by several microanalyses (examples in Figure 3).

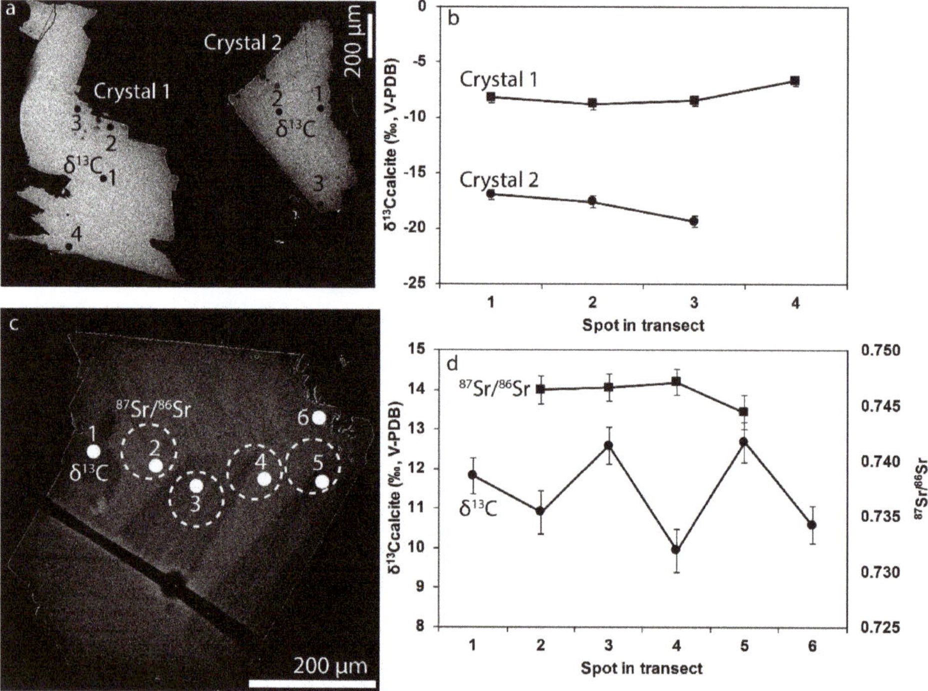

Figure 2. Transects of $\delta^{13}C$ and $^{87}Sr/^{86}Sr$ values from microanalyses within polished calcite crystals mounted in epoxy. (**a**) Back-scattered electron (BSE) image of two crystals from sample 181 m with $\delta^{13}C$ for each crystal transect shown in (**b**). (**c**) BSE image of a crystal from sample 185 m with $\delta^{13}C$ and $^{87}Sr/^{86}Sr$ values shown in (**d**). The crystal in (**c**) is more heterogeneous in $\delta^{13}C$ and BSE intensity than crystals in (**a**), due to higher degree of growth zonation in the former. Note the larger spot size of the Sr isotope analysis (80 μm, spot craters shown in Figure 4) compared to C isotope analysis (10 μm). Error bars represent 2σ.

Table 1. SIMS-analyses of $\delta^{34}S$ in chalcopyrite in sample 185 m. Sample and Trout Lake chalcopyrite reference material, in analytical sequence.

Crystal/Reference	^{32}S cps ($\times 10^9$)	$^{34}S/^{32}S$	$\pm$ abs	$\delta^{34}S_{CDT}$ (°)	$\pm\sigma$ (°)
		Drift Corrected			
Reference	0.66	0.0442764	0.0000017	0.36	0.06
Reference	0.68	0.0442713	0.0000023	0.24	0.07
1	0.66	0.0443742	0.0000021	2.57	0.07
1	0.66	0.0443723	0.0000018	2.52	0.07
1	0.65	0.0443817	0.0000030	2.73	0.09
2	0.66	0.0443697	0.0000026	2.46	0.08
2	0.66	0.0443756	0.0000018	2.60	0.07
3	0.65	0.0443733	0.0000019	2.55	0.07
Reference	0.66	0.0442731	0.0000013	0.28	0.06
3	0.66	0.0443553	0.0000031	2.14	0.09
3	0.66	0.0443692	0.0000023	2.45	0.07
3	0.66	0.0443696	0.0000020	2.46	0.07
3	0.66	0.0443794	0.0000029	2.68	0.08
4	0.66	0.0443826	0.0000045	2.76	0.11
4	0.66	0.0443832	0.0000020	2.77	0.07
Reference	0.66	0.0442755	0.0000018	0.34	0.07
4	0.66	0.0443779	0.0000025	2.65	0.08
5	0.65	0.0443602	0.0000021	2.25	0.07
6	0.65	0.0443216	0.0000023	1.38	0.07
7	0.66	0.0443781	0.0000026	2.65	0.08
8	0.66	0.0443895	0.0000026	2.91	0.08
8	0.66	0.0443752	0.0000024	2.59	0.07
Reference	0.66	0.0442745	0.0000027	0.31	0.08
9	0.65	0.0443958	0.0000019	3.05	0.07
9	0.65	0.0444002	0.0000023	3.15	0.07
9	0.66	0.0443685	0.0000030	2.44	0.08
9	0.66	0.0443857	0.0000017	2.83	0.06
10	0.66	0.0444065	0.0000030	3.30	0.09
10	0.65	0.0444176	0.0000021	3.55	0.07
Reference	0.64	0.0442718	0.0000022	0.25	0.07
10	0.64	0.0444125	0.0000028	3.43	0.08
10	0.63	0.0444023	0.0000035	3.20	0.09
11	0.63	0.0443028	0.0000034	0.95	0.09
11	0.64	0.0442781	0.0000025	0.39	0.08
11	0.64	0.0444033	0.0000022	3.22	0.07
12	0.64	0.0443509	0.0000027	2.04	0.08
Reference	0.64	0.0442744	0.0000020	0.31	0.07
13	0.63	0.0443912	0.0000024	2.95	0.07
13	0.64	0.0444193	0.0000022	3.59	0.07
14	0.63	0.0443902	0.0000027	2.93	0.08
14	0.64	0.0443839	0.0000021	2.79	0.07
14	0.63	0.0443679	0.0000025	2.42	0.08
14	0.66	0.0443547	0.0000032	2.12	0.09
15	0.64	0.0443929	0.0000042	2.99	0.11
Reference	0.65	0.0442740	0.0000026	0.30	0.08
Reference	0.65	0.0442776	0.0000023	0.38	0.07
Reference	0.65	0.0442705	0.0000024	0.22	0.07

Table 2. SIMS and laser ablation multi-collector inductively coupled plasma mass spectrometry (LA-MC-ICP-MS) data for calcite. SIMS-analyses of $\delta^{13}C$ in calcite in samples 181 and 185 m and LA-MC-ICP-MS-analyses of $^{87}Sr/^{86}Sr$ in sample 185 m. Sample and reference material, in analytical sequence (SIMS). Reference materials for $^{87}Sr/^{86}Sr$ are listed in table below. SD=Standard deviation, SE=Standard error.

Crystal/Reference	^{12}C cps [× 10^9]	$^{13}C/^{12}C$ (Drift Corrected)	±abs (‰)	$\delta^{13}C_{PDB}$ (‰)	±σ (‰)	Sr isotopes (Sampling Time)	$^{87}Sr/^{86}Sr$ [1]	2SD [2]	$^{84}Sr/^{86}Sr$	2SE	$^{87}Rb/^{86}Sr$ [3]	2SE	Total Sr-Beam
Reference	0.0179	0.0108741	0.0000025	0.20	0.45								
Reference	0.0181	0.0108747	0.0000030	0.26	0.48								
Reference	0.0180	0.0108753	0.0000025	0.31	0.45								
Reference	0.0180	0.0108720	0.0000025	0.01	0.45								
181 m, crystal 1	0.0184	0.0107828	0.0000030	−8.20	0.47								
181 m, crystal 1	0.0187	0.0107763	0.0000027	−8.79	0.46								
181 m, crystal 1	0.0184	0.0107798	0.0000032	−8.47	0.48								
Reference	0.0180	0.0108795	0.0000025	0.70	0.45								
181 m, crystal 1	0.0184	0.0107996	0.0000025	−6.65	0.45								
181 m, crystal 2	0.0184	0.0106878	0.0000030	−16.93	0.47								
181 m, crystal 2	0.0186	0.0106803	0.0000025	−17.62	0.44								
181 m, crystal 2	0.0176	0.0106614	0.0000032	−19.37	0.48								
185 m, crystal 1	0.0185	0.0110004	0.0000025	11.82	0.45								
185 m, crystal 1	0.0184	0.0109904	0.0000041	10.90	0.54	31.0	0.7464	0.0013	0.0580	0.0038	0.00015	0.00047	0.167
Reference	0.0179	0.0108689	0.0000037	−0.27	0.51								
185 m, crystal 1	0.0186	0.0110087	0.0000028	12.58	0.47	30.6	0.7466	0.0012	0.0602	0.0047	<DL		0.140
185 m, crystal 1	0.0184	0.0109799	0.0000042	9.93	0.55	31.0	0.7471	0.0012	0.0548	0.0035	0.00014	0.00028	0.238
185 m, crystal 1	0.0183	0.0110098	0.0000034	12.68	0.50	31.0	0.7444	0.0015	0.0578	0.0047	0.00010	0.00051	0.144
185 m, crystal 1	0.0187	0.0109870	0.0000027	10.58	0.46								
Reference	0.0179	0.0108738	0.0000028	0.17	0.47								
Reference	0.0179	0.0108801	0.0000029	0.75	0.47								
Reference	0.0179	0.0108717	0.0000025	−0.02	0.45								
Reference	0.0176	0.0108796	0.0000031	0.71	0.48								
Reference	0.0177	0.0108739	0.0000028	0.19	0.46								
Reference	0.0174	0.0108713	0.0000026	−0.05	0.45								

Table 2. *Cont.*

Measurements of Sr isotope reference materials
Primary reference material: 'Ecnomiosa gerda'

Spot Number	Sampling Time	$^{87}Sr/^{86}Sr^1$	$2SD^2$	$^{84}Sr/^{86}Sr$	2SE	$^{87}Rb/^{86}Sr^3$	2SE	Total Sr-Beam
1	30.8	0.70914	0.00019	0.05629	0.00017	0.000055	0.000025	3.26
2	30.4	0.70923	0.00015	0.05657	0.00018	0.000047	0.000017	3.77
3	30.4	0.70906	0.00015	0.05646	0.00017	0.000007	0.000024	4.02
4	30.8	0.70926	0.00016	0.05656	0.00014	0.000043	0.000016	4.36
5	30.4	0.70914	0.00016	0.05646	0.00014	0.000055	0.000016	4.04
6	30.8	0.70921	0.00015	0.05658	0.00017	0.000035	0.000016	4.10
7	31.2	0.70913	0.00017	0.05656	0.00016	0.000031	0.000021	3.61
8	30.4	0.70913	0.00019	0.05646	0.00013	0.000009	0.000015	4.22
9	30.4	0.70915	0.00024	0.05642	0.00021	0.000015	0.000025	2.81
10	30.4	0.70908	0.00021	0.05639	0.00020	0.000036	0.000021	3.29
11	30.8	0.70923	0.00017	0.05650	0.00018	0.000043	0.000018	3.55
12	30.4	0.70918	0.00018	0.05640	0.00019	0.000047	0.000019	3.51
13	30.8	0.70923	0.00016	0.05643	0.00013	0.000065	0.000017	3.96
Average		0.70917		0.05647				
2 SD		0.00012		0.00017				

Secondary reference material: modern oyster shell from Western Australia

Spot Number	Sampling Time	$^{87}Sr/^{86}Sr^1$	$2SD^2$	$^{84}Sr/^{86}Sr$	2SE	$^{87}Rb/^{86}Sr^3$	2SE	Total Sr-Beam
1	29.2	0.70915	0.00019	0.05666	0.00032	0.000050	0.000036	2.23
2	29.2	0.70932	0.00019	0.05664	0.00024	0.000079	0.000025	2.50
3	29.6	0.70927	0.00020	0.05616	0.00026	0.000109	0.000030	2.50
Average		0.70925		0.05649				
2 SD		0.00017		0.00057				

Modern seawater: $^{87}Sr/^{86}Sr$ = 0.7091792 ± 0.0000021 (Mokadem et al. (2015)) [1] normalised to reference value for primary reference material 'Ecnomiosa gerda' (TIMS: $^{87}Sr/^{86}Sr$ = 0.709181 ± 0.000004, Kiel et al., (2014)). [2] propagated 2SD from repeated reference material measurements. [3] ^{87}Rb was calculated from ^{85}Rb (Rb-Factor = 0.3861), <DL = below detection limit.

Table 3. LA-ICP-MS in situ Rb-Sr data of reference materials.

		Primary ^{87}Rb/^{86}Sr Reference Material: BCR-2G							
Spot No.	Spot Size (µm)	^{87}Rb/^{86}Sr	1s error (%)	^{87}Sr/^{86}Sr	1s Error (%)	ρ	Age Error	Ratio RSE	Signal Length (sec)
6	50	0.3917	0.75%	0.70217	0.27%	0.18	-	0.67%	65
25	50	0.3994	0.64%	0.70428	0.32%	0.23	-	0.81%	65
26	50	0.3966	0.62%	0.70494	0.22%	0.14	-	0.54%	65
47	50	0.3941	0.58%	0.70713	0.30%	0.20	-	0.74%	65
48	50	0.3865	0.67%	0.70684	0.28%	0.44	-	0.73%	65
72	50	0.3920	0.53%	0.70481	0.28%	0.28	-	0.68%	65
73	50	0.3841	0.61%	0.70376	0.29%	0.23	-	0.72%	65
95	50	0.3935	0.62%	0.70451	0.27%	0.17	-	0.64%	65
96	50	0.3895	0.53%	0.70536	0.25%	0.35	-	0.63%	65
119	50	0.3858	0.64%	0.70474	0.22%	0.32	-	0.59%	65
120	50	0.3914	0.70%	0.70619	0.31%	0.29	-	0.78%	65
146	50	0.3884	0.52%	0.70788	0.25%	0.16	-	0.61%	65
147	50	0.3933	0.59%	0.70478	0.31%	0.19	-	0.74%	65
170	50	0.3886	0.81%	0.70568	0.38%	0.39	-	1.00%	50
171	50	0.3870	0.57%	0.70518	0.28%	0.17	-	0.66%	65
191	50	0.3906	0.60%	0.70494	0.27%	0.28	-	0.70%	65
192	50	0.3896	0.60%	0.70518	0.26%	0.23	-	0.66%	65
213	50	0.3860	0.67%	0.70455	0.28%	0.34	-	0.77%	65
214	50	0.3819	0.65%	0.70204	0.33%	0.44	-	0.85%	65
Average		0.39		0.70500					
SD		1.12%		0.21%					

		Primary ^{87}Sr/^{86}Sr Reference Material: NIST-SRM-610							
Spot No.	Spot Size (µm)	^{87}Rb/^{86}Sr	1s Error (%)	^{87}Sr/^{86}Sr	1s Error (%)	ρ	Age Error	Ratio RSE	Signal Length (sec)
1	50	2.3244	0.36%	0.70436	0.28%	0.41	-	0.81%	65
2	50	2.3246	0.44%	0.70910	0.23%	0.37	-	0.69%	65
3	50	2.3651	0.38%	0.71060	0.28%	0.32	-	0.82%	65
4	50	2.3469	0.42%	0.71208	0.25%	0.25	-	0.73%	65
23	50	2.3611	0.41%	0.71341	0.27%	0.24	-	0.75%	65
24	50	2.3586	0.41%	0.71120	0.34%	0.36	-	1.00%	65
45	50	2.3320	0.35%	0.70624	0.26%	0.20	-	0.75%	65
46	50	2.3416	0.42%	0.70980	0.32%	0.33	-	0.90%	65
70	50	2.3375	0.68%	0.70592	0.33%	0.45	-	0.95%	65
71	50	2.3407	0.57%	0.71322	0.32%	0.39	-	0.92%	65
93	50	2.3685	0.51%	0.71297	0.37%	0.31	-	1.10%	65
94	50	2.3635	0.63%	0.70978	0.27%	0.26	-	0.77%	55
117	50	2.3586	0.41%	0.70905	0.27%	0.32	-	0.81%	65
118	50	2.3488	0.45%	0.71169	0.24%	0.28	-	0.69%	65

Table 3. *Cont.*

Primary $^{87}Sr/^{86}Sr$ Reference Material: NIST-SRM-610									
Spot No.	Spot Size (μm)	$^{87}Rb/^{86}Sr$	1s Error (%)	$^{87}Sr/^{86}Sr$	1s Error (%)	ρ	Age Error	Ratio RSE	Signal Length (sec)
144	50	2.3456	0.51%	0.71069	0.32%	0.29	-	0.92%	65
145	50	2.3474	0.48%	0.71028	0.23%	0.33	-	0.70%	65
168	50	2.3556	0.53%	0.70719	0.27%	0.28	-	0.79%	65
169	50	2.3328	0.38%	0.71201	0.29%	0.33	-	0.83%	65
189	50	2.3042	0.43%	0.70903	0.25%	0.43	-	0.74%	65
190	50	2.2924	0.40%	0.70751	0.25%	0.33	-	0.74%	65
211	50	2.2880	0.42%	0.71029	0.23%	0.19	-	0.64%	65
212	50	2.2683	0.55%	0.70698	0.30%	0.36	-	0.87%	40
Average		2.3366		0.7097					
SD		1.16%		0.35%					

Secondary Reference Material: La Posta											
Spot No.	Spot Size (μm)	Mineral	$^{87}Rb/^{86}Sr$	1s error (%)	$^{87}Sr/^{86}Sr$	1s error (%)	ρ	Age Error	Ratio RSE	Signal Length (sec)	Status	Note
15	50	Biotite	849	4.32%	1.83	4.26%	0.94	3.63%	13.15%	55		
16	50	Biotite	1025	5.23%	2.03	5.03%	0.94	4.29%	29.62%	55		
51	50	Biotite	891	21.45%	1.97	21.22%	0.95	9.81%	22.48%	10	rejected	short signal
52	50	Biotite	250	5.01%	1.03	4.06%	0.67	12.02%	12.49%	30		
53	50	Biotite	261	4.93%	1.06	4.12%	0.76	12.45%	15.29%	30		
99	50	Biotite	35.2	10.92%	0.764	1.37%	0.78	50.36%	3.86%	65	rejected	unstable ablation, high $^{87}Rb/^{86}Sr$ error
100	50	Biotite	261	3.54%	1.06	2.73%	0.76	6.07%	6.75%	58		
123	50	Biotite	512	4.34%	1.39	3.56%	0.87	4.91%	11.94%	60		
124	50	Biotite	858	4.61%	1.83	4.31%	0.92	3.60%	11.24%	67		
150	50	Biotite	140	5.14%	0.896	2.49%	0.65	12.29%	8.81%	37		
151	50	Biotite	185	4.11%	0.947	3.19%	0.75	17.51%	15.68%	60		
203	50	Biotite	53.4	14.51%	0.904	4.84%	0.25	39.23%	11.82%	8	rejected	short signal
204	50	Biotite	19.1	24.36%	0.792	3.31%	0.13	59.06%	8.89%	15	rejected	unstable ablation, high $^{87}Rb/^{86}Sr$ error

Initial $^{87}Sr/^{86}Sr$: 0.70483 ± 0.00014 [37]; Isochron age: 94.2 ± 2.9 Ma, MSWD=0.101.

Table 4. LA-ICP-MS in situ Rb-Sr data of samples 216m and 185m. nd = not detected.

Sample: 216m

Spot No.	Spot Size (μm)	Mineral	^{87}Rb/^{86}Sr	1s error (%)	^{87}Sr/^{86}Sr	1s error (%)	ρ	Age Error	Ratio RSE	Ablation Length (sec)	Status	Note
1	50	K-feldspar	17.0	0.70%	1.06	0.49%	0.54	1.25%	1.41%	65		
2	50	K-feldspar	17.9	1.68%	1.06	0.72%	0.75	1.40%	2.02%	65		
3	50	K-feldspar	20.4	1.63%	1.12	0.79%	0.51	1.87%	2.42%	25		
4	50	K-feldspar	15.6	1.31%	1.03	0.53%	0.69	1.22%	1.62%	65		
5	50	K-feldspar	16.8	0.94%	1.05	0.51%	0.45	1.33%	1.46%	50		
6	50	K-feldspar	13.8	0.73%	1.00	0.72%	0.67	2.03%	2.07%	35		
7	50	K-feldspar	17.3	0.86%	1.07	0.63%	0.64	1.43%	1.91%	40		
8	50	K-feldspar	15.0	0.64%	1.02	0.67%	0.52	1.91%	1.96%	30		
9	50	K-feldspar	13.0	0.91%	0.981	0.53%	0.31	1.88%	1.57%	40		
10	50	K-feldspar	14.3	0.71%	1.00	0.47%	0.63	1.24%	1.32%	60		
11	50	K-feldspar	11.7	0.77%	0.962	0.54%	0.57	1.74%	1.58%	40		
12	50	K-feldspar	14.7	0.90%	1.00	0.53%	0.55	1.52%	1.63%	55		
13	50	K-feldspar	15.8	1.52%	1.03	0.64%	0.67	1.49%	1.86%	65		
14	50	K-feldspar	32.9	2.43%	1.29	1.06%	0.79	1.50%	3.13%	30	rejected	unstable ablation, age mix?
15	50	K-feldspar	37.0	62.90%	n.d.	n.d.	3.99	88.04%	66.99%	5	rejected	short ablation
16	50	K-feldspar	30.1	25.84%	1.32	36.36%	−1.06	78.49%	63.24%	5	rejected	short ablation
17	50	Albite	3.01	0.87%	0.771	0.39%	0.27	12.49%	1.11%	30		mineral sample from 185m
18	50	Albite	3.59	0.81%	0.781	0.30%	0.19	6.87%	0.87%	55		mineral sample from 185m
19	50	Albite	4.19	0.93%	0.783	0.39%	0.25	8.07%	1.13%	40		mineral sample from 185m
20	50	Albite	1.16	15.86%	0.773	1.61%	−0.11	47.10%	16.29%	65	rejected	high ^{87}Rb/^{86}Sr errors
21	50	Albite	3.72	7.93%	0.805	2.32%	0.47	35.30%	6.16%	30	rejected	high ^{87}Rb/^{86}Sr errors
22	50	Albite	3.10	8.69%	0.796	1.77%	0.28	47.17%	4.36%	30	rejected	high ^{87}Rb/^{86}Sr errors

Sample: 185m

Spot No.	Spot Size (μm)	Mineral	^{87}Rb/^{86}Sr	1s Error (%)	^{87}Sr/^{86}Sr	1s Error (%)	ρ	Age Error	Ratio RSE	Ablation Length (sec)	Status	Note
23	50	K-feldspar	241	1.24%	1.93	1.21%	0.87	1.01%	3.62%	65		
24	50	K-feldspar	263	1.92%	2.06	1.69%	0.86	1.36%	5.33%	50		
25	50	K-feldspar	158	1.43%	1.57	0.99%	0.82	1.03%	2.90%	60		
26	50	K-feldspar	203	2.22%	1.73	1.74%	0.93	1.35%	4.94%	60		
27	50	K-feldspar	212	1.51%	1.83	1.34%	0.87	1.19%	3.93%	60		
28	50	K-feldspar	222	1.41%	1.84	1.08%	0.81	1.06%	3.15%	65		
29	50	K-feldspar	169	1.34%	1.61	1.20%	0.85	1.24%	3.42%	60		
30	50	K-feldspar	166	1.37%	1.57	1.19%	0.80	1.37%	3.60%	60		
31	50	K-feldspar	169	1.52%	1.63	1.59%	0.75	1.85%	4.31%	60		
32	50	K-feldspar	546	2.60%	3.17	2.46%	0.95	1.07%	7.34%	60	rejected	unstable ablation, high errors
33	50	K-feldspar	182	1.78%	1.66	1.36%	0.90	1.15%	4.11%	60		
34	50	Calcite	-	-	0.752	0.68%	0.17	-	-	55		
35	50	Calcite	-	-	0.745	1.26%	−0.39	-	-	15		
36	50	Calcite	-	-	0.759	2.24%	−0.11	-	-	25		

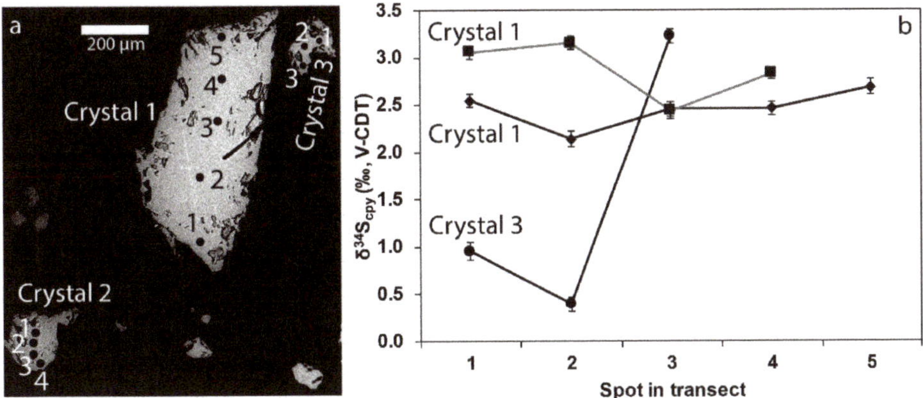

Figure 3. Transects of δ^{34}S values in polished chalcopyrite crystals. (**a**) BSE image with analytical spots marked. (**b**) δ^{34}S values of three crystals.

5.2. Rb/Sr Dating

The isochron constructed from 10 LA-ICP-MS analyses of adularia and three LA-ICP-MS analyses in addition to four LA-MC-ICP-MS analyses of calcite for sample 185 m yield an age of 356.6 ± 6.7 Ma (Figure 4, Table 4). Adularia crystals from sample 216 m paired with albite acquired in the 185 m fracture yield an age of 1457 ± 63 Ma (Appendix A Figure A1), but as this age represents fracture formation at more than 1 Gyr before the Lockne impact structure formed, we do not discuss it in more detail. It is, nevertheless, an important and expected observation that shows that pre-existing fractures were reactivated during the impact.

6. Discussion

The isotope record of the carbonate and sulfide crystals serves as an archive for processes in the fracture system over long time periods [19,20,43,44]. In the following sections the various (mostly microbial) processes that can be responsible for the observed isotope signatures are discussed. It is important to emphasize that the isotope composition of the minerals may represent the result of one or several (cryptic) processes in the fracture system. These processes (when biological) are usually distinguished by the kinetic isotope effect that occurs when microbial populations utilize a substrate (i.e., they alter the isotope composition due to preferential utilization of a specific isotope). In addition, the source of the C and S compounds and processes occurring during transport can also influence the isotope composition before the element is incorporated into a mineral. Several of the potential source compounds may also overlap in isotope composition making certain diagnostic determinations about processes and sources difficult when using isotope composition of minerals only. Abiotic fractionation can also overlap in magnitude with microbial fractionation, which may inhibit certain conclusions about microbial processes in the fracture system. However, some magnitudes of isotope fractionation and systematics have been used extensively as diagnostic tracers for specific microbial processes. These include C isotope markers for methanogenesis and methane oxidation and S isotope markers for microbial sulfate reduction [45–48].

6.1. Ancient Methanogenesis in the Fracture Voids

Methane is usually ^{13}C-depleted compared to other carbon compounds [49], especially when the methane is microbial in origin. As a consequence of the fractionation occurring during methanogenesis, which discriminates against ^{13}C, the residual CO_2 becomes ^{13}C-rich [50]. Subsequent involvement of the residual C into precipitating carbonate minerals is therefore a useful diagnostic C-isotope tracer for methanogenesis, for instance by ^{13}C-rich secondary carbonates in sedimentary basins [51] or in fractured crystalline rocks in the Fennoscandian shield [24,43,52,53]. The significantly ^{13}C-enriched calcite in the Lockne sample 185 m ($\delta^{13}C_{calcite}$ values as heavy as +12.7‰, Figure 2d) is thus a strong line of evidence for microbial methanogenesis *in situ* in the deep fracture system especially since potential abiotic methane-forming chemical processes such as serpentinization, graphite metamorphism or Fischer-Tropsch type reactions [54,55] are unlikely under the local physico-chemical and geological conditions. Strong ^{13}C-enrichment in carbonate is typically associated with microbial methane formed by carbonate reduction, as opposed to acetate (methyle-type) fermentation that involves smaller kinetic carbon isotope effects [56]. Heavy $\delta^{13}C_{CO2}$ values are particularly typical for secondary microbial methane formed from thermogenic precursors [56], which may involve a previous methane oxidation step that can occur anaerobically, if associated with microbial reduction of e.g., sulfate [56]. Sulfide minerals can form as a result of microbial sulfate reduction (MSR) and the chalcopyrite observed in sample 185 m may, therefore, have formed due to MSR in a step preceding methanogenesis. The MSR metabolism involves a kinetic isotope effect that discriminates against [34]S resulting in production of a typically strongly [34]S-depleted hydrogen sulfide [57]. As there is no significant S isotope fractionation during sulfide precipitation the $\delta^{34}S_{sulfide}$ composition serves as a diagnostic marker for MSR, because abiotic sulfate reduction at higher temperatures involves smaller fractionation [58]. However, the $\delta^{34}S_{chalcopyrite}$ values determined by SIMS (+0.4 to 3.6‰) are not very depleted. The initial $\delta^{34}S$ composition of the sulfate in the system is unknown, which makes estimates of the fractionation difficult. If we anticipate that the source sulfate had a similar composition as Paleozoic seawater (+15 to +35‰ [59]) an isotope enrichment of 10–35‰ would be needed to produce the detected $\delta^{34}S_{chalcopyrite}$ values. This degree of isotope enrichment and uncertainty of initial sulfate composition, inhibit any certain interpretation of the $\delta^{34}S_{chalcopyrite}$ values as microbial in origin, because they cannot be fully distinguished from thermochemical reduction, and furthermore are overlapping with hydrothermal sulfide of magmatic origin [58,60,61]. Rayleigh isotope reservoir effects may also have occurred in the fracture system, which can modify the S isotope signature significantly [20].

The overall low $\delta^{13}C_{calcite}$ values of the 181 m sample are more in line with an organic origin, such as plants or oil/petroleum/kerogen/bitumen [62,63], and may reflect microbial utilization (oxidation) of such carbon sources, for instance coupled with sulfate reduction. The crystal-specific $\delta^{13}C$ variance in this sample can be due to temporal variation of the microbial processes and substrates in the fracture and/or due to spatial micro-scale variability within the fracture voids, as reported in other deep fractured bedrock systems in the Fennoscandian Shield [19,24]. Organic compounds may have descended from overlying, but presently completely eroded Paleozoic sedimentary successions that once covered the Fennoscandian shield [64]. Thermochronological studies show that the sedimentary successions had considerable thickness in the mid-Paleozoic era [65]. Infiltration of bitumen and other organic compounds of surficial origin into the crystalline bedrock occurred presumably from thermally heated organic-rich shales in the lower parts of the sedimentary pile [19,66] of the Fennoscandian shield, and elsewhere (e.g., UK and Australia [67,68]). Putative bitumen occurrences have also been reported from the Lockne fracture system [33], and may thus have been influential for the deep biosphere community in the deep fracture system.

6.2. New Age Constraints for the Fracture Assemblage in the Impact Structure and Microbial Communities

The age of the feldspar-calcite assemblage in sample 185 m of 356.6 ± 6.7 Ma (Figure 4) is about 100 Myr younger than the estimated age of the impact [28]. The fracture system and mineral paragenesis of the impact structure have previously been considered to be of the same age as the impact [33], and consequently the fossilized microbial communities (dominantly fungi) observed in the fracture system have been assigned the same age [16]. Our new radiometric age determination requires an updated model for the timing of the fracture system activation in the impact structure, and consequently also for the preserved microbial remains within this system. The obtained age, at the Devonian–Carboniferous transition, is in line with secondary mineral assemblages in fractures elsewhere in the Fennoscandian shield, i.e., feldspar-calcite veins, with fossilized microorganisms at 300 m depth at Forsmark, Sweden, dated to between 355 ± 14 and 402 ± 10 Ma, and at Oskarshamn dated to 358 ± 12 and 393 ± 15 Ma by *in situ* Rb/Sr geochronology [19–21]. At Forsmark, a few calcite crystals showed methanogenesis-related ^{13}C-rich composition, and/or ^{13}C-depleted values typical for anaerobic oxidation of methane [19,21]. This suggests that a regional fracture-activation event at these times enabled fluid circulation, which induced microbial activity and secondary mineral formation in the bedrock. Fluids from overlying organic-rich shales were at this event allowed to descend into the deeper crystalline bedrock fracture network, such as the deeply fragmented bedrock at Lockne and thereby provided substrates to the otherwise energy-poor deep biosphere.

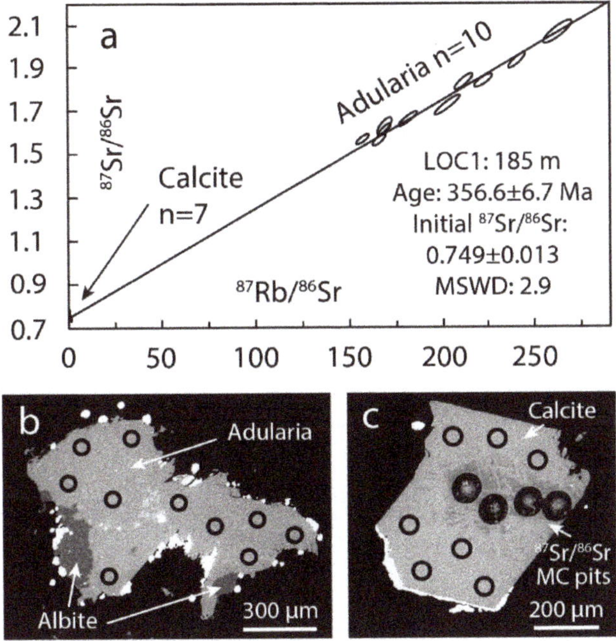

Figure 4. In situ Rb-Sr dating results of the 185m sample with (**a**) the isochron diagram and data, and (**b,c**) individual spots marked as black circles on BSE images of the analysed adularia and calcite mineral grains.

Fluid inclusions of secondary calcite from the current drill core have indicated fluids with a salinity around 20 eq. wt% $CaCl_2$ corresponding to brine, and homogenization temperatures in the range of 77–218 °C [33], which at least for the salinities correspond to fluid inclusions in similar fracture coating calcite in Oskarshamn and Forsmark in Sweden, from southern Finland, and to Caledonian mineralizations [19,53,69–71]. The δ^{18}O values of calcite and quartz [33,72] at Lockne are also in line with compositions of the fracture-filling calcite reported from other parts of the Fennoscandian shield [19,53,69–71,73]. Most of the fluid inclusion homogenization temperatures in the previous study [33] were in the 100–180 °C range, which is generally too high for microbial activity. The large temperature span of 77–218 °C, however, does suggest that fluid temperatures dropped at later stages of the fluid circulation event in the fractures. During the early part of the fluid circulation when temperatures were >100 °C, K-feldspar-albite-calcite-quartz formed and thermally mature bitumen was probably transported from overlying sediments, because asphaltite can become viscous and mobilized at elevated temperatures [66]. The occurrence of methane in the fluid inclusions in calcite of this stage also marks mobility of thermogenic hydrocarbons from the shale source [33] into the deeper fracture system. When temperatures decreased after the initial phase of hydrothermal circulation, deep subsurface microbial communities could use the bitumen as an energy source by oxidizing the organic matter and subsequently producing secondary methane as discussed above, as well as sustaining heterotrophic eukaryotic activity in the form of fungi. This late stage calcite formation is in line with brine type fluid inclusions with lower homogenization temperatures, down to 77 °C, which is more suitable for microbial colonization, or alternatively a group of calcite crystals in the 50–70 °C span [33].

The revised radiometric ages of the fracture re-activation set the fungal colonization in a different environmental context. At the time of the impact, the area was covered by a shallow sea [33], and it was believed that marine fungi were introduced into the impact-generated hydrothermal system by ingress of seawater through the fractured oceanic crust [16]. Similarly, a seawater origin is suggested for biological material responsible for sulfur reduction in the Rochechouart impact crater in France [74]. However, 356 Myr ago the current area of the Fennoscandian shield hosting the Lockne crater was not marine but continental [31], which indicates that the fungi are not marine but represent fungi belonging to the continental deep biosphere at that time. Although the brine-type fluid inclusions of the calcites may, at first glance, point to a marine origin, it is relevant to consider that deep continental brines are result of several different processes. In addition to marine infiltration, these processes include prolonged water-rock interaction, mixing of fluids with different sources, concentration during freezing and descent of sedimentary brines [75–78]. The latter has importance in Canada where this process has been studied extensively [79]. There, it has been suggested that deep continental brines result from infiltration of saline water to the crystalline basement aquifer from basinal sedimentary brines and evaporites when the shield was covered by sedimentary successions in the Paleozoic [79]. This means that brines may develop under continental conditions if marine sedimentary successions overlay the shield.

Contemporaneous methanogenesis is in agreement with previous fossilized fungal-prokaryotic communities in granitic deep settings [23], and suggest a potential synergy. Methanogenic activity and formation of sulfides indicates anoxic conditions in the system and explain the pristine carbonaceous nature of the fungal fossils and lack of mineralization by clays and Fe-oxides, which is common among fungal fossilization in oxygenated deep environments and dominates the findings in the oceanic crust [80].

Although impact craters and associated impact-generated hydrothermal systems likely are favourable environments for microbial life [1,2] owing to their increased pore space for endolithic communities, and heat convection favourable for ecosystems [5], our revised model for the ancient Lockne microbial community shows that colonization took place long after the impact, in contrast to previous models. The role of the impact-induced hydrothermal system is, therefore, negligible for the fossilized community observed in the fractures. Instead, reactivation of the fractures occurred 100 Myr after the impact event, in a post-Caledonian extension event documented previously throughout the Fennoscandian shield. Fractures and pore space opened as a result of the impact thus likely provided pathways for the extension-related fluids. However, the extent of reutilization is likely to be local given that the relatively small amounts of melt rocks are scattered along the crater rims, implying that small-scale hydrothermal cells did not produce extensive wall-rock dissolution. This extension correlates with a heating episode in the crystalline basement of southern Sweden that relates to burial by Caledonian foreland sediments [65], whereas Caledonian nappe structures overthrusted the Lockne area during the Silurian and early Devonian [81]. Tectonothermal or thermometamorphic overprinting has caused resetting of ^{40}Ar/^{39}Ar melt rock ages in several impact structures, including Gardnos in Norway [82], Acraman in Australia [83] and Charlevoix in Canada [84]. Following rapid uplift after Caledonian orogenic collapse, thermochronology records temperatures of 250–300 °C at 350 Ma [85] towards western Norway where the metamorphic temperatures were the highest during the orogeny as well as in northwestern Norway [86]. In the nappes immediately west to the Lockne area, cooling rates of 15 °C/Myr brought the rocks through 475 °C at 425 Ma [87]. Similar rapid cooling episodes at 350 Ma in western Norway have been attributed to extension concurrent with rapid uplift [88].

Despite the fact that the Rb-Sr isotopic system is more resistant to heating than Ar systems, the lack of thermochronological constraints in the Lockne area at 350 Ma means that temperatures high enough to disturb the radiogenic chronometer cannot be ruled out. However, the hydrothermal minerals of Lockne neither feature any traces of such an overprint in terms of diffusion or dissolution textures nor radiogenic isotopic inhomogeneity or disturbances on spatial and depth scales. Furthermore, no signs of severe degradation of the fossils are observed, which would certainly be expected if experiencing a regional thermal event. Instead, the pristine fungi structure indicates precipitation through extension-facilitated hydrothermal activity where bituminous material from overlying shales was introduced into the fracture system and provided energy for microbial activity. In that sense, the majority of the studied hydrothermal mineralizations are not impact induced and thus not of astrobiological importance in a Martian context as was previously suggested [16].

7. Conclusions

The age of previously described fungal fossils from the Lockne crater (458 Ma), Sweden, is here revised. In situ Rb/Sr dating of secondary calcite-feldspar (356.6 ± 6.7 Ma) shows that the fungal colonization is not associated with the impact-induced hydrothermal system but took place at least 100 Myr after the impact event. This revised age excludes the previous notion that marine fungi were introduced by seawater recharge, and instead suggests that the fungi may have been established in the deep continental crust underneath remnant marine successions already at 356 Ma or shortly after. Microscale stable isotope data of ^{13}C-enriched calcite further suggest methanogenesis occurred in the fracture system, potentially in synergy with the fungi.

Author Contributions: M.I. M.T. H.D. performed the SEM-analyses, M.T. performed the LA-ICP-MS Rb–Sr dating, data reduction and interpretation, H.D. performed the SIMS and LA-MC-ICP analyses together with M.J.W. and E.K.+M.S., respectively. M.T. wrote the paper together with H.D. and M.I.

Funding: This research was funded by Swedish research council (contract 2017-05186 to H.D., 2017-04129 to M.I.), Formas (contract 2017-00766 to H.D. and M.W.), and a Villum Investigator Grant to Don Canfield (No. 16518).

Acknowledgments: Thanks to Erik Sturkell for access to drill cores. We acknowledge NordSIM for provision of facilities and experimental support under Swedish Research Council grant no 2014-06375. K. Lindén, and Johan Hogmalm are thanked for analytical or sample preparation assistance. This is NordSIM publication #603 and Vegacenter publication #017.

Conflicts of Interest: The authors declare no conflict of interest and the founding sponsors had no role in the design of the study; in the collection, analyses, or interpretation of data; in the writing of the manuscript, and in the decision to publish the results.

Appendix A

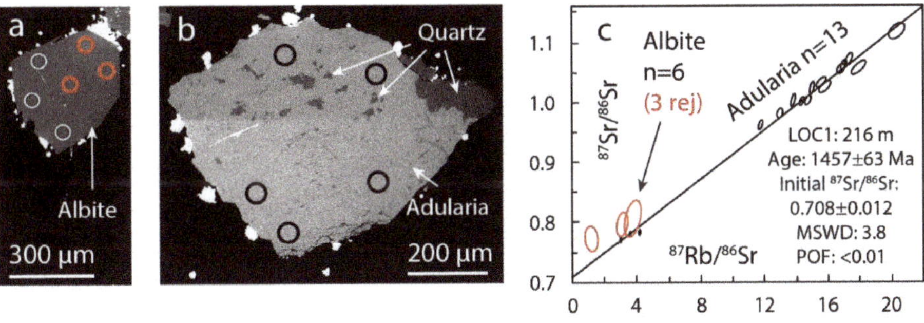

Figure A1. In situ Rb-Sr dating results of the 216 m sample with (**a**) albite spots as white circles, rejected albite spots as red circles, (**b**) adularia spots in one of three grains as black circles on BSE images and (**c**) the isochron diagram and the resulting data.

References

1. Versh, E.; Kirsimäe, K.; Jõeleht, A. Development of potential ecological niches in impact-induced hydrothermal systems: The small-to-medium size impacts. *Planet. Space Sci.* **2006**, *54*, 1567–1574. [CrossRef]
2. Osinski, G.R.; Tornabene, L.L.; Banerjee, N.R.; Cockell, C.S.; Flemming, R.; Izawa, M.R.; McCutcheon, J.; Parnell, J.; Preston, L.J.; Pickersgill, A.E.; et al. Impact-generated hydrothermal systems on Earth and Mars. *Icarus* **2013**, *224*, 347–363. [CrossRef]
3. Kring, D.A. Impact events and their effect on the origin, evolution, and distribution of life. *GSA today* **2000**, *10*, 1–7.
4. Cockell, C.S. The origin and emergence of life under impact bombardment. *Philos. Trans. Soc. B: Boil. Sci.* **2006**, *361*, 1845–1856. [CrossRef] [PubMed]
5. Cockell, C.S.; Lee, P. The biology of impact craters – a review. *Boil. Rev.* **2002**, *77*, 279–310.
6. Naumov, M.V. Principal features of impact-generated hydrothermal circulation systems: Mineralogical and geochemical evidence. *Geofluids* **2002**, *5*, 165–184. [CrossRef]
7. Ames, D.E.; Watkinson, D.H.; Parrish, R.R. Dating of a regional hydrothermal system induced by the 1850 Ma Sudbury impact event. *Geology* **1998**, *26*, 447. [CrossRef]
8. Jõeleht, A.; Kirsimäe, K.; Plado, J.; Versh, E.; Ivanov, B. Cooling of the Kärdla impact crater: II. Impact and geothermal modeling. *Meteor. Planet. Sci.* **2005**, *40*, 21–33. [CrossRef]
9. Arp, G.; Kolepka, C.; Simon, K.; Karius, V.; Nolte, N.; Hansen, B.T. New evidence for persistent impact-generated hydrothermal activity in the Miocene Ries impact structure, Germany. *Meteorit. Planet. Sci.* **2013**, *48*, 2491–2516. [CrossRef]
10. Schmieder, M.; Jourdan, F. The Lappajärvi impact structure (Finland): Age, duration of crater cooling, and implications for early life. *Geochim. Cosmochim. Acta* **2013**, *112*, 321–339. [CrossRef]
11. Kenny, G.G.; Schmieder, M.; Whitehouse, M.J.; Nemchin, A.A.; Morales, L.F.; Buchner, E.; Bellucci, J.J.; Snape, J.F. A new U-Pb age for shock-recrystallised zircon from the Lappajärvi impact crater, Finland, and implications for the accurate dating of impact events. *Geochim. Cosmochim. Acta* **2019**, *245*, 479–494. [CrossRef]
12. Parnell, J.; Boyce, A.; Thackrey, S.; Muirhead, D.; Lindgren, P.; Mason, C.; Taylor, C.; Still, J.; Bowden, S.; Osinski, G.R.; et al. Sulfur isotope signatures for rapid colonization of an impact crater by thermophilic microbes. *Geology* **2010**, *38*, 271–274. [CrossRef]
13. Sapers, H.M.; Osinski, G.R.; Banerjee, N.R.; Preston, L.J. Enigmatic tubular features in impact glass. *Geology* **2014**, *42*, 471–474. [CrossRef]
14. Lindgren, P.; Ivarsson, M.; Neubeck, A.; Broman, C.; Henkel, H.; Holm, N.G. Putative fossil life in a hydrothermal system of the Dellen impact structure, Sweden. *Int. J. Astrobiol.* **2010**, *9*, 137–146. [CrossRef]
15. Hode, T.; Cady, S.L.; von Dalwigk, I.; Kristiansson, P. Evidence of Ancient Microbial Life in an Impact Structure and Its Implications for Astrobiology. In *From Fossils to Astrobiology: Records of Life on Earth and Search for Extraterrestrial Biosignatures*; Seckbach, J., Walsh, M., Eds.; Springer Netherlands: Dordrecht, The Netherlands, 2008; pp. 249–273.
16. Ivarsson, M.; Broman, C.; Sturkell, E.; Ormö, J.; Siljeström, S.; Van Zuilen, M.; Bengtson, S. Fungal colonization of an Ordovician impact-induced hydrothermal system. *Sci. Rep.* **2013**, *3*, 3487. [CrossRef]
17. Ivarsson, M.; Holm, N.G.; Neubeck, A. The Deep Biosphere of the Subseafloor Igneous Crust. In *Trace Metal Biogeochemistry and Ecology of Deep-Sea Hydrothermal Vent Systems*; Demina, L.L., Galkin, V.S., Eds.; Springer International Publishing: Cham, Switzerland, 2016; pp. 143–166.
18. Ivarsson, M.; Bengtson, S.; Skogby, H.; Lazor, P.; Broman, C.; Belivanova, V.; Marone, F. A Fungal-Prokaryotic Consortium at the Basalt-Zeolite Interface in Subseafloor Igneous Crust. *PLOS ONE* **2015**, *10*, e0140106. [CrossRef] [PubMed]
19. Drake, H.; Heim, C.; Roberts, N.M.; Zack, T.; Tillberg, M.; Broman, C.; Ivarsson, M.; Whitehouse, M.J.; Åström, M.E. Isotopic evidence for microbial production and consumption of methane in the upper continental crust throughout the Phanerozoic eon. *Earth Planet. Sci. Lett.* **2017**, *470*, 108–118. [CrossRef]
20. Drake, H.; Whitehouse, M.J.; Heim, C.; Reiners, P.W.; Tillberg, M.; Hogmalm, K.J.; Dopson, M.; Broman, C.; Åström, M.E. Unprecedented 34 S-enrichment of pyrite formed following microbial sulfate reduction in fractured crystalline rocks. *Geobiology* **2018**, *16*, 556–574. [CrossRef] [PubMed]

21. Drake, H.; Ivarsson, M.; Tillberg, M.; Whitehouse, M.J.; Kooijman, E. Ancient Microbial Activity in Deep Hydraulically Conductive Fracture Zones within the Forsmark Target Area for Geological Nuclear Waste Disposal, Sweden. *Geosciences* **2018**, *8*, 211. [CrossRef]

22. Tillberg, M.; Drake, H.; Zack, T.; Hogmalm, J.; Åström, M. In Situ Rb-Sr Dating of Fine-grained Vein Mineralizations Using LA-ICP-MS. *Procedia Earth Planet. Sci.* **2017**, *17*, 464–467. [CrossRef]

23. Drake, H.; Ivarsson, M.; Bengtson, S.; Heim, C.; Siljeström, S.; Whitehouse, M.J.; Broman, C.; Belivanova, V.; Åström, M.E. Anaerobic consortia of fungi and sulfate reducing bacteria in deep granite fractures. *Nat. Commun.* **2017**, *8*, 55. [CrossRef]

24. Drake, H.; Åström, M.E.; Heim, C.; Broman, C.; Åström, J.; Whitehouse, M.; Ivarsson, M.; Siljestrom, S.; Sjövall, P. Extreme 13C depletion of carbonates formed during oxidation of biogenic methane in fractured granite. *Nat. Commun.* **2015**, *6*, 7020. [CrossRef]

25. Drake, H.; Tullborg, E.-L.; Whitehouse, M.; Sandberg, B.; Blomfeldt, T.; Åström, M.E. Extreme fractionation and micro-scale variation of sulphur isotopes during bacterial sulphate reduction in deep groundwater systems. *Geochim. Cosmochim. Acta* **2015**, *161*, 1–18. [CrossRef]

26. Drake, H.; Åström, M.E.; Tullborg, E.-L.; Whitehouse, M.; Fallick, A.E. Variability of sulphur isotope ratios in pyrite and dissolved sulphate in granitoid fractures down to 1km depth – Evidence for widespread activity of sulphur reducing bacteria. *Geochim. Cosmochim. Acta* **2013**, *102*, 143–161. [CrossRef]

27. Drake, H.; Ivarsson, M. The role of anaerobic fungi in fundamental biogeochemical cycles in the deep biosphere. *Fungal Boil. Rev.* **2018**, *32*, 20–25. [CrossRef]

28. Alwmark, C.; Schmitz, B. Extraterrestrial chromite in the resurge deposits of the early Late Ordovician Lockne crater, central Sweden. *Earth Planet. Sci. Lett.* **2007**, *253*, 291–303. [CrossRef]

29. Shuvalov, V.; Ormö, J.; Lindström, M. Hydrocode Simulation of the Lockne Marine Target Impact Event. In *Impact Tectonics*; Koeberl, C., Henkel, H., Eds.; Springer: Berlin/Heidelberg, Germany, 2005; pp. 405–422.

30. Högdahl, K. 1.86–1.85 Ga intrusive ages of K-feldspar megacryst-bearing granites in the type area of the Revsund granites in Jämtland County, central Sweden. *GFF* **2000**, *122*, 359–366. [CrossRef]

31. Lindström, M.; Lundqvist, J.; Lundqvist, T. *Sveriges Geologi Från Urtid till Nutid*; Studentlitteratur: Lund, Sweden, 2000.

32. Lindström, M.; Sturkell, E.F. Geology of the Early Palaeozoic Lockne impact structure, Central Sweden. *Tectonophysics* **1992**, *216*, 169–185. [CrossRef]

33. Sturkell, E.F.; Broman, C.; Forsberg, P.; Torssander, P. Impact-related hydrothermal activity in the Lockne impact structure, Jämtland, Sweden. *Eur. J. Miner.* **1998**, *10*, 589–606. [CrossRef]

34. Crowe, D.E.; Vaughan, R.G. Characterization and use of isotopically homogeneous standards for in situ laser microprobe analysis of 34 S/ 32 S ratios. *Am. Miner.* **1996**, *81*, 187–193. [CrossRef]

35. Kiel, S.; Glodny, J.; Birgel, D.; Bulot, L.G.; Campbell, K.A.; Gaillard, C.; Graziano, R.; Kaim, A.; Lazăr, I.; Sandy, M.R.; et al. The Paleoecology, Habitats, and Stratigraphic Range of the Enigmatic Cretaceous Brachiopod Peregrinella. *PLOS ONE* **2014**, *9*, e109260. [CrossRef] [PubMed]

36. Mokadem, F.; Parkinson, I.J.; Hathorne, E.C.; Anand, P.; Allen, J.T.; Burton, K.W. High-precision radiogenic strontium isotope measurements of the modern and glacial ocean: Limits on glacial–interglacial variations in continental weathering. *Earth Planet. Sci. Lett.* **2015**, *415*, 111–120. [CrossRef]

37. Zack, T.; Hogmalm, K.J. Laser ablation Rb/Sr dating by online chemical separation of Rb and Sr in an oxygen-filled reaction cell. *Chem. Geol.* **2016**, *437*, 120–133. [CrossRef]

38. Hogmalm, K.J.; Zack, T.; Karlsson, A.K.-O.; Sjöqvist, A.S.L.; Garbe-Schönberg, D. In situ Rb–Sr and K–Ca dating by LA-ICP-MS/MS: an evaluation of N 2 O and SF 6 as reaction gases. *J. Anal. At. Spectrom.* **2017**, *32*, 305–313. [CrossRef]

39. Nebel, O.; Mezger, K.; Scherer, E.; Münker, C. High precision determinations of 87Rb/85Rb in geologic materials by MC-ICP-MS. *Int. J. Mass Spectrom.* **2005**, *246*, 10–18. [CrossRef]

40. Jochum, K.P.; Weis, U.; Stoll, B.; Kuzmin, D.; Yang, Q.; Raczek, I.; Jacob, D.E.; Stracke, A.; Birbaum, K.; Frick, D.A.; et al. Determination of Reference Values for NIST SRM 610-617 Glasses Following ISO Guidelines. *Geostand. Geoanalytical* **2011**, *35*, 397–429. [CrossRef]

41. Elburg, M.; Vroon, P.; Van Der Wagt, B.; Tchalikian, A. Sr and Pb isotopic composition of five USGS glasses (BHVO-2G, BIR-1G, BCR-2G, TB-1G, NKT-1G). *Chem. Geol.* **2005**, *223*, 196–207. [CrossRef]

42. Villa, I.; De Bièvre, P.; Holden, N.; Renne, P. IUPAC-IUGS recommendation on the half life of 87 Rb. *Geochim. Cosmochim. Acta* **2015**, *164*, 382–385. [CrossRef]

43. Sahlstedt, E.; Karhu, J.A.; Pitkänen, P.; Whitehouse, M. Biogenic processes in crystalline bedrock fractures indicated by carbon isotope signatures of secondary calcite. *Appl. Geochem.* **2016**, *67*, 30–41. [CrossRef]

44. Sahlstedt, E.; Karhu, J.; Pitkanen, P.; Whitehouse, M. Implications of sulfur isotope fractionation in fracture-filling sulfides in crystalline bedrock, Olkiluoto, Finland. *Appl. Geochem.* **2013**, *32*, 52–69. [CrossRef]

45. Knittel, K.; Boetius, A. Anaerobic Oxidation of Methane: Progress with an Unknown Process. *Annu. Microbiol.* **2009**, *63*, 311–334. [CrossRef] [PubMed]

46. Peckmann, J.; Thiel, V. Carbon cycling at ancient methane–seeps. *Chem. Geol.* **2004**, *205*, 443–467. [CrossRef]

47. Canfield, D.E.; Olesen, C.A.; Cox, R.P. Temperature and its control of isotope fractionation by a sulphate-reducing bacterium. *Geochim. Cosmochim. Acta* **2006**, *70*, 548–561. [CrossRef]

48. Johnston, D.T.; Wing, B.A.; Farquhar, J.; Kaufman, A.J.; Strauss, H.; Lyons, T.W.; Kah, L.C.; Canfield, D.E. Active Microbial Sulfur Disproportionation in the Mesoproterozoic. *Science* **2005**, *310*, 1477–1479. [CrossRef] [PubMed]

49. Whiticar, M.J. Carbon and hydrogen isotope systematics of bacterial formation and oxidation of methane. *Chem. Geol.* **1999**, *161*, 291–314. [CrossRef]

50. Boehme, S.E.; Blair, N.E.; Chanton, J.P.; Martens, C.S. A mass balance of 13C and 12C in an organic-rich methane-producing marine sediment. *Geochim. Cosmochim. Acta* **1996**, *60*, 3835–3848. [CrossRef]

51. Budai, J.M.; Walter, L.M.; Ku, T.C.W.; Martini, A.M. Fracture-fill calcite as a record of microbial methanogenesis and fluid migration: a case study from the Devonian Antrim Shale, Michigan Basin. *Geofluids* **2002**, *2*, 163–183. [CrossRef]

52. Drake, H.; Tullborg, E.-L.; Hogmalm, K.J.; Åström, M.E. Trace metal distribution and isotope variations in low-temperature calcite and groundwater in granitoid fractures down to 1km depth. *Geochim. Cosmochim. Acta* **2012**, *84*, 217–238. [CrossRef]

53. Sandström, B.; Tullborg, E.-L. Episodic fluid migration in the Fennoscandian Shield recorded by stable isotopes, rare earth elements and fluid inclusions in fracture minerals at Forsmark, Sweden. *Chem. Geol.* **2009**, *266*, 126–142. [CrossRef]

54. Kietäväinen, R.; Purkamo, L. The origin, source, and cycling of methane in deep crystalline rock biosphere. *Front. Microbiol.* **2015**, *6*, 725. [CrossRef]

55. Etiope, G.; Lollar, B.S. ABIOTIC METHANE ON EARTH. *Rev. Geophys.* **2013**, *51*, 276–299. [CrossRef]

56. Pallasser, R. Recognising biodegradation in gas/oil accumulations through the δ 13 C compositions of gas components. *Org. Geochem.* **2000**, *31*, 1363–1373. [CrossRef]

57. Sim, M.S.; Bosak, T.; Ono, S. Large sulfur isotope fractionation does not require disproportionation. *Science* **2011**, *333*, 74–77. [CrossRef] [PubMed]

58. Machel, H.G.; Krouse, H.R.; Sassen, R. Products and distinguishing criteria of bacterial and thermochemical sulfate reduction. *Appl. Geochem.* **1995**, *10*, 373–389. [CrossRef]

59. Strauss, H. The isotopic composition of sedimentary sulfur through time. *Palaeogeogr. Palaeoclim. Palaeoecol.* **1997**, *132*, 97–118. [CrossRef]

60. Field, C.W.; Fifarek, R.H. Light stable-isotopes systematics in the epithermal environment. *Rev. Econ. Geology* **1985**, *2*, 99–128.

61. Kiyosu, Y.; Krouse, H.R. The role of organic acid in the abiogenic reduction of sulfate and the sulfur isotope effect. *Geochem. J.* **1990**, *24*, 21–27. [CrossRef]

62. Vlierboom, F.; Collini, B.; Zumberge, J. The occurrence of petroleum in sedimentary rocks of the meteor impact crater at Lake Siljan, Sweden. *Org. Geochem.* **1986**, *10*, 153–161. [CrossRef]

63. Roberts, H.H.; Aharon, P. Hydrocarbon-derived carbonate buildups of the northern Gulf of Mexico continental slope: A review of submersible investigations. *Geo-Marine Lett.* **1994**, *14*, 135–148. [CrossRef]

64. Koistinen, T.; Stephens, M.B.; Bogatchev, V.; Nordgulen, Ø.; Wennerström, M.; Korkonen, J. *Geological Map of the Fennoscandian Shield Scale 1:2,000,000*; Geological Surveys of Finland, Norway and Sweden and the Northwest Department of Natural Resources of Russia: Espoo, Finland, 2001.

65. Guenthner, W.R.; Reiners, P.W.; Drake, H.; Tillberg, M. Zircon, titanite, and apatite (U-Th)/He ages and age-eU correlations from the Fennoscandian Shield, southern Sweden. *Tectonics* **2017**, *36*, 1254–1274. [CrossRef]

66. Sandström, B.; Tullborg, E.-L.; De Torres, T.; Ortiz, J.E. The occurrence and potential origin of asphaltite in bedrock fractures, Forsmark, central Sweden. *GFF* **2006**, *128*, 233–242. [CrossRef]

67. Parnell, J.; Baba, M.; Bowden, S.; Muirhead, D. Subsurface biodegradation of crude oil in a fractured basement reservoir, Shropshire, UK. *J. Geol. Soc.* **2017**, *174*, 655–666. [CrossRef]

68. McCready, A.J.; Stumpfl, E.F.; Melcher, F. U/Th-rich bitumen in Archean granites and Palaeoproterozoic metasediments, Rum Jungle Mineral Field, Australia: implications for mineralizing fluids. *Geofluids* **2003**, *3*, 147–159. [CrossRef]

69. Drake, H.; Tullborg, E.-L. Paleohydrogeological events recorded by stable isotopes, fluid inclusions and trace elements in fracture minerals in crystalline rock, Simpevarp area, SE Sweden. *Appl. Geochem.* **2009**, *24*, 715–732. [CrossRef]

70. Sahlstedt, E.; Karhu, J.A.; Pitkänen, P. Indications for the past redox environments in deep groundwaters from the isotopic composition of carbon and oxygen in fracture calcite, Olkiluoto, SW Finland. *Isot. Environ. Heal. Stud.* **2010**, *46*, 370–391. [CrossRef]

71. Alm, E.; Sundblad, K.; Huhma, H. *Sm-Nd Isotope Determinations of Low-Temperature Fluorite-Calcite-Galena Mineralization in the Margins of the Fennoscandian Shield*; Report R-05-66; Svensk Kärnbränslehantering AB: Stockholm, Sweden, 2005; p. 58.

72. Broman, C.; Sturkell, E.; Fallick, A.E. Oxygen isotopes and implications for the cavity-grown quartz crystals in the Lockne impact structure, Sweden. *GFF* **2011**, *133*, 101–107. [CrossRef]

73. Maskenskaya, O.M.; Drake, H.; Broman, C.; Hogmalm, J.K.; Czuppon, G.; Astrom, M.E. Source and character of syntaxial hydrothermal calcite veins in Paleoproterozoic crystalline rocks revealed by fine-scale investigations. *Geofluids* **2014**, *14*, 495–511. [CrossRef]

74. Simpson, S.; Boyce, A.; Lambert, P.; Lindgren, P.; Lee, M.; Simpson, S.; Boyce, A. Evidence for an impact-induced biosphere from the δ 34 S signature of sulphides in the Rochechouart impact structure, France. *Earth Planet. Sci. Lett.* **2017**, *460*, 192–200. [CrossRef]

75. Bottomley, D.; Katz, A.; Chan, L.; Starinsky, A.; Douglas, M.; Clark, I.; Raven, K. The origin and evolution of Canadian Shield brines: evaporation or freezing of seawater? New lithium isotope and geochemical evidence from the Slave craton. *Chem. Geol.* **1999**, *155*, 295–320. [CrossRef]

76. Lüders, V.; Rickers, K. Fluid inclusion evidence for impact-related hydrothermal fluid and hydrocarbon migration in Creataceous sediments of the ICDP-Chicxulub drill core Yax-1. *Meteorit. Planet. Sci.* **2004**, *39*, 1187–1197. [CrossRef]

77. Zurcher, L.; Kring, D.A.; Barton, M.D.; Dettman, D.; Rollog, M. Stable isotope record of post-impact fluid activity in the Yaxcopoil-1 borehole, Chicxulub impact structure, Mexico T. In *Large Meteorite Impacts III*; Kenkmann, T., Hörz, F., Deutsch, A., Eds.; Geological Society of America: Boulder, CO, USA, 2005; pp. 223–238.

78. Katz, A.; Starinsky, A.; Marion, G.M. Saline waters in basement rocks of the Kaapvaal Craton, South Africa. *Chem. Geol.* **2011**, *289*, 163–170. [CrossRef]

79. Gascoyne, M. Hydrogeochemistry, groundwater ages and sources of salts in a granitic batholith on the Canadian Shield, southeastern Manitoba. *Appl. Geochem.* **2004**, *19*, 519–560. [CrossRef]

80. Ivarsson, M.; Bengtson, S.; Drake, H.; Francis, W. Fungi in Deep Subsurface Environments. In *Advances in Applied Microbiology*; Academic Press: Cambridge, MA, USA, 2017.

81. Ladenberger, A.; Be'eri-Shlevin, Y.; Claesson, S.; Gee, D.G.; Majka, J.; Romanova, I.V. Tectonometamorphic evolution of the Åreskutan Nappe—Caledonian history revealed by SIMS U–Pb zircon geochronology. *Geol. Soc. London Publ.* **2014**, *390*, 337–368. [CrossRef]

82. Grier, J.A.; Swindle, T.D.; Kring, D.A.; Melosh, H.J. Argon-40/argon-39 analyses of samples from the Gardnos impact structure, Norway. *Meteorit. Planet. Sci.* **1999**, *34*, 803–807. [CrossRef]

83. Schmieder, M.; Tohver, E.; Jourdan, F.; Denyszyn, S.; Haines, P. Zircons from the Acraman impact melt rock (South Australia): Shock metamorphism, U–Pb and 40 Ar/ 39 Ar systematics, and implications for the isotopic dating of impact events. *Geochim. Cosmochim. Acta* **2015**, *161*, 71–100. [CrossRef]

84. Buchner, E.; Schmieder, M.; Schwarz, W.H.; Trieloff, M.; Hopp, J.; Spray, J.G. Dating the Charlevoix Impact Structure (Québec, Canada) – A Tough Nut to Crack in 40Ar/39Ar Geochronology. In Proceedings of the 41st Lunar and Planetary Science Conference, The Woodlands, TX, USA, 1–5 March 2010; p. 41.

85. Ksienzyk, A.K.; Wemmer, K.; Jacobs, J.; Fossen, H.; Schomberg, A.C.; Süssenberger, A.; Lünsdorf, N.K.; Bastesen, E. Post-Caledonian brittle deformation in the Bergen area, West Norway: Results from K–Ar illite fault gouge dating. *Nor. J. Geol.* **2016**, *96*, 275–299. [CrossRef]

86. Steltenpohl, M.G.; Carter, B.T.; Andresen, A.; Zeltner, D.L. 40Ar/39Ar Thermochronology of Late- and Postorogenic Extension in the Caledonides of North-Central Norway. *J. Geol.* **2009**, *117*, 399–414. [CrossRef]

87. Hacker, B.R.; Gans, P.B. Continental collisions and the creation of ultrahigh-pressure terranes: Petrology and thermochronology of nappes in the central Scandinavian Caledonides. *GSA Bull.* **2005**, *117*, 117. [CrossRef]
88. Eide, E.A.; Torsvik, T.H.; Andersen, T.B.; Arnaud, N.O. Early Carboniferous Unroofing in Western Norway: A Tale of Alkali Feldspar Thermochronology. *J. Geol.* **1999**, *107*, 353–374. [CrossRef]

Review

The Carbon-Isotope Record of the Sub-Seafloor Biosphere

Patrick Meister [1],* and Carolina Reyes [2]

[1]   Department of Geodynamics and Sedimentology, University of Vienna, Althanstr. 14, 1090 Vienna, Austria
[2]   Department of Environmental Geosciences, University of Vienna, Althanstr. 14, 1090 Vienna, Austria; creyes6@gmail.com
*   Correspondence: patrick.meister@univie.ac.at

Received: 13 November 2019; Accepted: 29 November 2019; Published: 5 December 2019

Abstract: Sub-seafloor microbial environments exhibit large carbon-isotope fractionation effects as a result of microbial enzymatic reactions. Isotopically light, dissolved inorganic carbon (DIC) derived from organic carbon is commonly released into the interstitial water due to microbial dissimilatory processes prevailing in the sub-surface biosphere. Much stronger carbon-isotope fractionation occurs, however, during methanogenesis, whereby methane is depleted in ^{13}C and, by mass balance, DIC is enriched in ^{13}C, such that isotopic distributions are predominantly influenced by microbial metabolisms involving methane. Methane metabolisms are essentially mediated through a single enzymatic pathway in both *Archaea* and *Bacteria*, the Wood–Ljungdahl (WL) pathway, but it remains unclear where in the pathway carbon-isotope fractionation occurs. While it is generally assumed that fractionation arises from kinetic effects of enzymatic reactions, it has recently been suggested that partial carbon-isotope equilibration occurs within the pathway of anaerobic methane oxidation. Equilibrium fractionation might also occur during methanogenesis, as the isotopic difference between DIC and methane is commonly on the order of 75‰, which is near the thermodynamic equilibrium. The isotopic signature in DIC and methane highly varies in marine porewaters, reflecting the distribution of different microbial metabolisms contributing to DIC. If carbon isotopes are preserved in diagenetic carbonates, they may provide a powerful biosignature for the conditions in the deep biosphere, specifically in proximity to the sulphate–methane transition zone. Large variations in isotopic signatures in diagenetic archives have been found that document dramatic changes in sub-seafloor biosphere activity over geological time scales. We present a brief overview on carbon isotopes, including microbial fractionation mechanisms, transport effects, preservation in diagenetic carbonate archives, and their implications for the past sub-seafloor biosphere and its role in the global carbon cycle. We discuss open questions and future potentials of carbon isotopes as archives to trace the deep biosphere through time.

Keywords: carbon isotopes; deep biosphere; diagenetic carbonates; methanogenesis; anaerobic methane oxidation; Wood–Ljungdahl pathway

---

1. Introduction

Carbon, in its reduced form, is not only the essential building material of life, but due to its large isotope variations, it can also serve as a tracer of biogeochemical processes in the environment and as an indicator of the state of the global carbon cycle. During assimilation of CO_2 to organic matter in the water column, carbon is depleted in ^{13}C by 20–30‰; however, variations in δ^{13}C in ocean and atmosphere are usually in the few-permil range. These variations are essentially balanced by input to and output from the ocean and atmosphere, and they only change upon variations in rates of primary production and burial of organic carbon relative to inorganic carbon (Broeker, 1970) [1].

In contrast, isotopic compositions of dissolved carbon species in marine sedimentary porewater show a large range of values. These systems, which are part of a deep biosphere, are exclusively inhabited by Prokaryotes, whereby the "deep subsurface biosphere" has been operationally defined as an "ecosystem that persists at least one metre, if not more" (Edwards et al., 2012) [2]. The deep biosphere is organized as zones of different metabolic activity in the sequence of downward decreasing redox potential (Froelich et al., 1979) [3], whereby the presence and extension of these zones vary considerably in different regions of the ocean (D'Hondt et al., 2004) [4]. Sulphate reduction represents the most abundant anaerobic respiration process, and, although dissimilatory degradation of organic matter generally exhibits small fractionation effects (e.g., Hayes et al., 1989) [5], it delivers dissolved inorganic carbon (DIC) strongly depleted in ^{13}C to the porewater. Furthermore, large fractionation effects are observed during fermentation reactions where the pools of CO_2 and CH_4 are involved. Biogenic methane often shows δ^{13}C values as negative as $-100‰$ relative to the Vienna Peedee Belemnite (VPDB) standard (Claypool and Kaplan, 1974) [6], while positive values of up to $+35‰$ have been observed in dissolved inorganic carbon (DIC) related to methane production (Heuer et al., 2009) [7]. In turn, strongly ^{13}C-depleted DIC is produced as a result of anaerobic oxidation of methane (AOM). Ultimately, the different metabolic processes result in a large variability in inorganic carbon, especially in proximity to the sulphate–methane transition zone.

Inorganic carbon can be incorporated in diagenetic carbonates and thereby become preserved as part of the mineral phase. Often, diagenetic carbonates form as a result of microbial dissimilatory activity, which also contributes to global carbon burial (Schrag et al., 2013) [8]. Diagenetic carbonates can trap the isotopic signature of the surrounding fluid from which they were precipitated, and preserve it for millions of years as part of the geological record. As opposed to carbon isotopes in marine carbonates, which vary in δ^{13}C in the few-permil range and can be used to reconstruct the global carbon cycle, diagenetic carbonates can show large variations based on type and rate of microbial activity and provide information on past biogeochemical conditions at a specific location. Indeed, previous studies have shown that carbon-isotope values preserved in diagenetic carbonates can be used as an archive of past deep biosphere activity (e.g., Kelts and McKenzie, 1984; Malone et al., 2002; Meister et al., 2007; Meister, 2015) [9–12]. Recent studies also showed that other light isotope systems (i.e., isotopes of light elements) in diagenetic phases, e.g., sulphur isotopes preserved in diagenetic pyrite (Meister et al., 2019a) [13], provide evidence similar to carbon isotopes, indicating that in the past, the conditions in the deep biosphere were different from today, and that biogeochemical zones migrated upwards and downwards in the sediment. Despite these insights, currently there are several problems that hamper a more detailed interpretation of diagenetic carbon-isotope records: (1) Fractionation effects are incompletely understood; (2) diffusive mixing and non-steady state conditions result in a complex mixture of isotopic compositions from different sources; and (3) models to quantitatively predict carbonate precipitation are not sufficiently developed.

Here, we provide a brief overview of the current state of knowledge of carbon-isotope effects related to the main anaerobic metabolic processes, sulphate reduction, methanogenesis, and AOM, occurring in marine sediments (Section 2). We discuss the different fractionation mechanisms, as part of enzymatic pathways, including a distinction between kinetic and potential equilibrium fractionation effects. In Section 3, we assess how different microbial processes affect carbon-isotope profiles in DIC and CH_4 in marine porewaters and how these profiles are subject to diffusive mixing and advective transport. Section 4 is focused on the controls on diagenetic carbonate formation in the sub-seafloor biosphere and how combined effects of carbon-isotope fractionation, transport, and mineral precipitation result in diagenetic carbon-isotope records. In Section 5, we discuss the importance of diagenetic carbonate burial on secular variation in the global carbon cycle and its ^{13}C signature and how we can trace these variations back in time to reconstruct the evolution of a dynamic sub-seafloor biosphere through Earth's history.

2. Carbon-Isotope Fractionation by Different Microbial Pathways

2.1. Kinetic Fractionation

Stable isotope fractionation strongly depends on the molecular pathways of metabolic reactions and is mostly the result of a kinetic effect, whereby a molecule having the same thermal vibration energy but higher mass is less likely to overcome the activation energy barrier of the reaction (Hoefs, 2018) [14]. Molecular pathways are mostly known in detail, but it remains unclear at which step of the pathway the isotope fraction occurs. Hence, from an isotopic point of view, many metabolic reactions represent a black box, and fractionation effects have been determined for the overall reactions.

In practice, two approaches have been used to determine microbial carbon-isotope fractionation effects: First, measuring carbon-isotope distributions in metabolites, such as CH_4, DIC, and other organic intermediates dissolved in natural porewaters can provide an "apparent" fractionation factor. Second, carbon-isotope fractionation factors have been determined directly from microcosm (Alperin et al., 1992) [15] or pure culture experiments. By convention, the fractionation factor α is defined as the isotope ratio of the reactant divided by the isotope ratio of the product. For convenience, fractionation can also be expressed by the separation factor, which is the difference in permil of the $\delta^{13}C$ of reactant and the product ($\varepsilon = (\alpha - 1) \cdot 1000$; Hayes, 1993; 2004) [16,17].

Sulphate reduction: Fractionation during dissimilatory sulphate reduction (Equation (1)) is considered insignificant (e.g., Claypool and Kaplan, 1974) [6].

$$2\,[CH_2O] + SO_4^{2-} \rightarrow HS^- + 2\,HCO_3^- + H^+ \tag{1}$$

Assuming that average organic matter $[CH_2O]$ is consumed quantitatively, while intermediate pools are very small, essentially no fractionation effects should be expected in natural sediments. In addition, fractionation factors determined in culture experiments using defined substrates for dissimilatory sulphate reduction (Londry and Des Marais, 2003) [18] were near to one.

Anaerobic methane oxidation: Whiticar and Faber (1986) and Alperin et al. (1988) [19,20] showed that a small kinetic fractionation ($\alpha = 1.0088$) occurs if sulphate reduction is linked to the anaerobic oxidation of methane (AOM; Equation (2)), resulting in a ^{13}C enrichment of the residual CH_4 pool.

$$CH_4 + SO_4^{2-} \rightarrow HS^- + HCO_3^- + H_2O \tag{2}$$

AOM culture experiments yielded significantly higher kinetic fractionation factors ($\alpha = 1.012\text{--}1.039$) than were evident from field observations (e.g., Holler et al., 2009) [21]. The reason for this discrepancy is unclear at present.

Methanogenesis: Fractionation effects during methanogenesis vary over a large range dependent on the enzymatic pathway and experimental conditions as shown in the histogram of separation factors compiled from the literature (Figure 1). During acetoclastic methanogenesis:

$$2\,CH_3COOH \rightarrow CO_2 + CH_4 \tag{3}$$

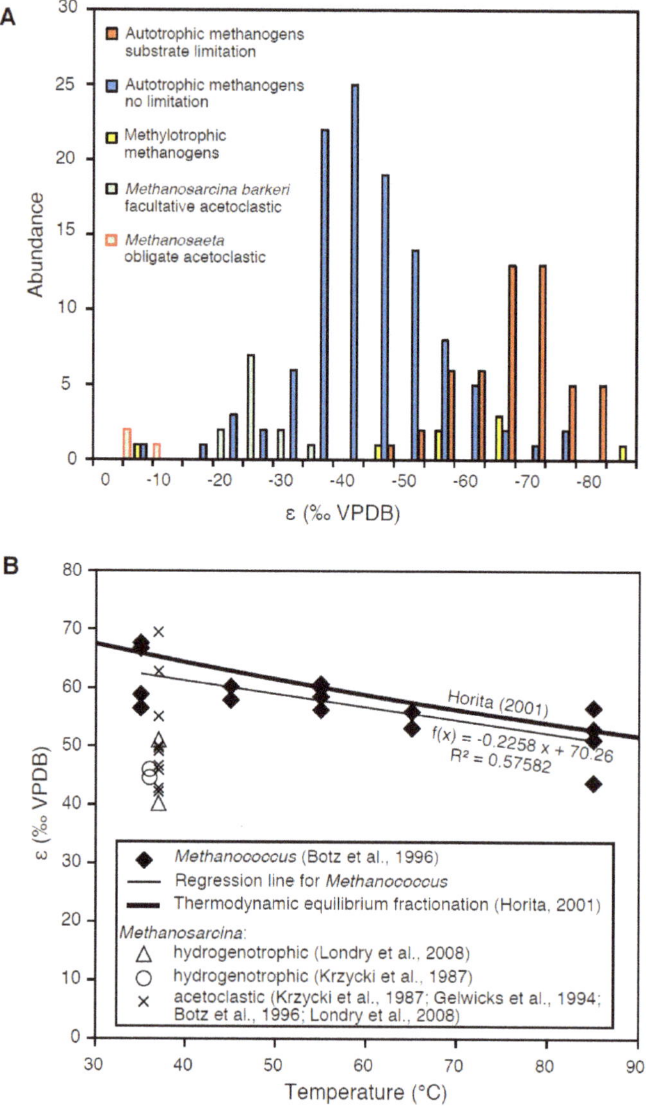

Figure 1. Compilation of separation factors ε from culture experiments using species of *Methanosarcina*, *Methanococcus*, *Methanobacterium*, *Methanothermobacter*, and *Methanothermococcus* grown under different conditions: (**A**) Histogram showing the abundance distribution of separation factors for methanogens grown with acetate, hydrogen, and methanol/trimethylamine. Distinctively larger isotope effects are observed during hydrogenotrophic growth, in particular, under substrate limitation, whereas minor fractionation occurs during acetoclastic growth. Thereby, obligate acetoclasts show insignificant fractionation. Data are compiled from Games et al. (1978) [22]; Fuchs et al. (1979) [23]; Belyaev et al. (1983) [24]; Balabane et al. (1987) [25]; Krzycki et al. (1987) [26]; Gelwicks et al. (1994) [27]; Botz et al. (1996) [28]; Summons et al., 1998 [29]; Valentine et al. (2004) [30]; Londry et al. (2008) [31]; Okumura et al., (2016) [32]; and Miller et al. (2018) [33]. (**B**) Temperature dependence of separation factors in experiments with *Methanococcus* growing under H_2 limitation (Botz et al., 1996) [28] in comparison to the thermodynamic isotope equilibrium from Horita (2001) [34].

which is a fermentation reaction, the observed fractionation factor between reactant (acetate) and product (CH_4) is small. While fractionation of the carboxyl carbon may be uncoupled from the methyl carbon, Sugimoto and Wada (1993) [35] showed in sediment incubation experiments that this reaction only produces a minor intramolecular isotopic difference. Pure-culture incubations with the facultative acetoclast *Methanosarcina barkeri* by Krzycki et al. (1987) [26], Gelwicks et al. (1994) [27], and Londry et al. (2008) [31] show a fractionation effect in the range of −20‰ to −30‰, in contrast to the obligate acetoclast *Methanosaeta thermophila*, showing insignificant fractionation (Valentine et al., 2004 [30]; Figure 1A).

Much larger fractionation effects were found for hydrogenotrophic methanogenesis (Whiticar et al., 1986) [36] with an apparent fractionation factor α = 1.05–1.09. However, it needs to be taken into account that for each mole of H_2 produced by fermentation of organic matter also 0.5 mol of CO_2 are produced, following the overall reaction:

$$2 \, [CH_2O] \xrightarrow{+2H_2O} 2 \, CO_2 + 4 \, H_2 \xrightarrow{-2H_2O} CO_2 + CH_4 \tag{4}$$

Therefore, the apparent separation factors determined in the field do not provide the exclusive fractionation factors of the pure methanogenic step but include the CO_2 from fermentation. Methanogenic culture experiments have been performed with *Methanosarcina barkeri* (Games et al., 1978 [22]; Krzycki et al., 1987 [26]; Londry et al., 2008 [31]), an acetoclastic methanogen capable of growing autotrophically, as well as with several strains of the obligate autotrophic methanogens *Methanococcus* (Botz et al., 1996) [28], *Methanobacterium* (Games et al., 1978 [22]; Fuchs et al., 1979 [23]; Belyaev et al., 1983 [24]; Balabane et al., 1987 [25]; Okumura et al., 2016 [32]; Miller et al., 2018 [33]), *Methanothermobacter* (Valentine et al., 2004 [30]; Okumura et al., 2016) [32], and *Methanothermococcus* (Okumura et al., 2016) [32] under conditions supplied with H_2 and CO_2. As shown by the histogram in Figure 1A, distinctly larger separation factors (ε) are observed for hydrogenotrophic than for acetoclastic methanogenesis, independent of the organism used. Also, a great range of separation factors is observed during methylotrophic growth, using *M. barkeri* or *Methanococcoides burtonii* supplied with methanol or trimethylamine (Summons et al., 1998 [29]; Londry et al., 2008 [31]; data included in Figure 1A).

Using a flow-through fermentor system, Botz et al. (1996) [28] were able to maintain growth of different species of *Methanococcus* under substrate limitation, which generally resulted in larger separation factors. A similar effect was achieved using a fermenting coculture, providing limited hydrogen substrate (Okumura et al., 2016 [32]), or under strongly alkaline conditions (Miller et al., 2018 [33]). Orange bars in Figure 1A highlight the significantly larger isotope effects under substrate-limited conditions. These conditions may better resemble the natural system, where often substrate limitation prevails and organisms grow extremely slowly (cf. "the starving majority"; Jørgensen and D'Hondt, 2006) [37].

2.2. Equilibrium Fractionation

Not entirely clear is the observation that methane often shows extremely negative $\delta^{13}C$ values at sulphate–methane transition zones (SMTZ), i.e., at the depth at which it is almost entirely consumed. If this were the result of kinetic fractionation, CH_4 would be expected to show less negative values (because light CH_4 is preferentially oxidized). Even though autotrophic methanogenesis may contribute to lower values, the isotope effect of AOM should dominate.

Recently, Holler et al. (2009, 2011) [21,38] and Yoshinaga et al. (2014) [39] published results of experimental studies using mixed cultures of sulphate-reducing bacteria and methanotrophic archaea, forming a consortium capable of anaerobically oxidizing methane. They suggest that the AOM reaction is reversible, as an amount of dissolved inorganic carbon (DIC) is channelled back into the methane pool. This was demonstrated by addition of ^{14}C-labelled DIC, which ended up in the methane during the experiment. This finding is remarkable since it provides insight into the enzymatic pathway. Yoshinaga et al. (2014) [39] suggested that a partial isotopic equilibration occurs within this process, and this isotopic fractionation could explain the formation of methane with strongly negative $\delta^{13}C_{CH4}$ values at the sulphate–methane transition (SMT) zones in marine sediments. In the study of Yoshinaga et al. (2014) [39], a kinetic fractionation model was used, where kinetic fractionation occurs in both forward and backward directions and a steady state is reached in a diffusion-limited system (cf. Hayes et al., 2004 [17]). Since isotopic equilibrium can only be reached if true chemical equilibrium occurs (Urey and Greiff, 1935) [40], it is assumed that also the AOM reaction under natural porewater conditions is close to thermodynamic equilibrium. This is indeed the case as the free energy yield of AOM is minimal and depends on the concentration levels of reactant and product. While abiotically, CO_2 does not exchange isotopes with CH_4 under Earth surface temperature (Giggenbach, 1982) [41], the equilibrium fractionation effect between CH_4 and CO_2 is extrapolated from high temperature (Richet et al., 1977 [42]; Horita, 2001 [34]). The separation factor is rather large, on the order of 70‰ at ambient temperature, and indeed, such a large isotopic difference is observed in marine porewaters.

The idea of equilibrium carbon-isotope fractionation is not entirely new. It was already suggested by Bottinga (1969) [43] for methanogenesis and is supported by culture experiments (Valentine et al., 2004 [30]; Penning et al., 2005 [44]; Moran et al., 2005 [45]; Takai et al., 2008 [46]). While experiments performed under different conditions show large scatter in the separation factors, the experiments under substrate limitation show significantly larger separation factors (Figure 1A). Plotting exclusively the values from experiments with *Methanococcus* by Botz et al. (1996) [28] against incubation temperature (Figure 1B), whereby the initial time steps were omitted, results in a regression line that fits very well to the thermodynamic equilibrium separation factor from Horita (2001) [34]. These experiments strongly suggest that also during methanogenesis, isotopic equilibration occurs.

2.3. Potential Fractionation Effects within the Molecular Pathway

While radiotracer experiments provide valuable information on fractionation mechanisms, these findings can now be considered to discuss fractionation effects within the actual biochemical pathways. We focus here on the Wood–Ljungdahl pathway (WL pathway; also known as acetyl coenzyme-A pathway), which is the essential pathway used by methanogens, acetogens, as well as methanotrophs. The WL pathway is at the centre of all methane metabolisms and, therefore, should be mainly responsible for carbon-isotope fractionation in natural aquatic sediments.

The Wood-Ljungdahl pathway: The backbone of the WL pathway is shown in Figure 2. It starts with CO_2 being first reduced to formate by formate-dehydrogenase (FDH). Alternatively, methylated substrates, such as methanol or methylamine, may be used. In *Archaea*, the formate is further reduced along a cascade of reduction steps to methyl-tetrahydromethanopterin (4H-MTP). In *Bacteria*, tetrahydrofolate is used instead of 4H-MTP. At the core of the WL pathway, acetyl-coenzyme-A synthase (ACS) combines a methyl group, containing a reduced carbon, with a carboxyl carbon when forming acetyl-coenzyme-A. The carboxyl carbon is supplied from CO_2 via a carbon monoxide-dehydrogenase (CODH), which is part of the ACS-CODH cluster (also called CO-methylating acetyl-CoA synthase; Adam et al. 2018 [47]). The pathway ends with the production of acetate or may continue from acetyl-CoA further to biosynthesis. In methanogens, the methyl group from methyl-4H-MTP is passed to coenzyme-M (CoM) and released as CH_4.

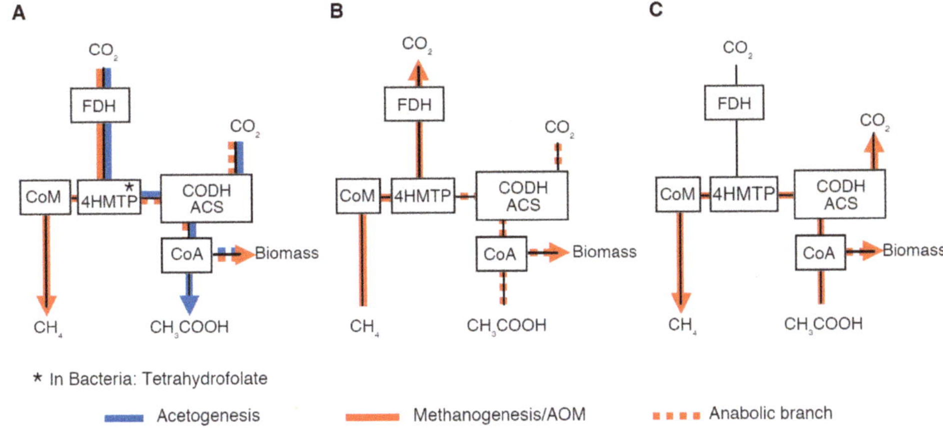

Figure 2. Wood–Ljungdahl pathway (black lines) and different directions in which metabolic reactions can run through the pathway: (**A**) Autotrophic methanogenesis (solid red line) and acetogenesis (blue line; via Tetrahydrofolate); (**B**) anaerobic methane oxidation; (**C**) acetoclastic methanogenesis. Also, the biosynthetic branch is indicated by dashed red lines. The scheme is drawn according to the WL pathway shown in Martin and Russel (2007) [48]. Abbreviations: Acetyl-CoA synthase: ACS; carbon monoxyde dehydrogenase: CODH; coenzyme-A: CoA; coenzyme-M: CoM; formate-dehydrogenase: FDH; tetrahydro-methanopterin: 4HMTP.

While the WL-pathway represents the most ancient pathway of carbon fixation that still exists in modern *Archaea* and *Bacteria* (cf. Weiss et al., 2016) [49], different organisms run the pathway in different directions, depending on the biogeochemical conditions, to gain energy and biomass. Figure 2 shows the three directions in which different organisms use the pathway. Autotrophic (i.e., hydrogenotrophic) methanogens such as *Methanococcus* (Figure 2A) use the MTP branch to reduce carbon and then pass the methyl group to CoM. However, they also use the ACS-CODH cluster for anabolism (dashed red line; Berghuis et al., 2019) [50]. Acetogens follow the same route (blue lines), but they do not produce CoM.

In the second case (Figure 2B), methanotrophs use CoM and the MTP-chain to oxidize CH_4 to CO_2 by running the pathway backwards. Nevertheless, for biosynthesis they still rely on carboxyl-carbon, which is directly fixed via the ACS-CODH cluster and, hence, not derived from methane. This has been confirmed by radiotracer experiments (Kellermann et al., 2012) [51], although isotopically light DIC that is delivered by AOM may be directly used within the community cluster (Alperin et al., 2009) [52]. Besides, the same pathway is also used by methanotrophic bacteria (Skennerton et al., 2017) [53].

Methanogens, using acetate as a substrate (acetoclastic methanogens, such as *Methanosarcina*), run the WL-pathway sideways (Figure 2C; see Thauer et al., 2008) [54]. They use the ACS-CODH cluster in reverse to cleave the carboxyl group from acetate and oxidize it to CO_2, while the methyl group is transferred to 4H-MTP and CoM.

Isotope fractionation in the WL-pathway: Having established the general pathways, we can now consider, where in the pathway the essential fractionation effect occurs. Valentine et al. (2004) [30] discuss the concept of differential reversibility within the WL pathway, where the hydrogenotrophic methanogenic steps from CO_2 to 4H-MTP are reversible but the methyl transfer to CoM represents a kinetic bottleneck (Figure 3). This step would only become reversible under strong substrate limitation, and indeed, isotopic equilibration seems to occur under such conditions as evidenced by separation factors of substrate-limited autotrophic (hydrogenotrophic) methanogens in Figure 1A,B. It is also conceivable that in the opposite direction, during AOM, the methyl transfer from CoM to 4H-MTP (Figure 3) would be the critical step, becoming reversible under electron acceptor limitation.

Meister et al. (2019b) [55] propose a model, where isotopic equilibration in fact occurs within the pathway, which would most likely be localized at this particular step.

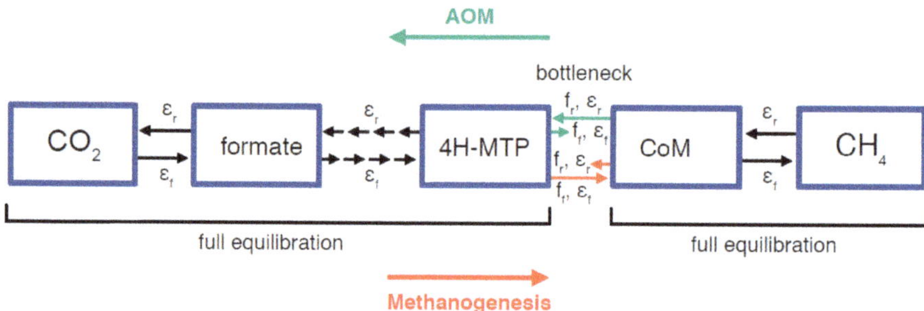

Figure 3. Scheme showing the differential reversibility of steps in the methanopterin-branch of the autotrophic methanogenic Wood–Ljungdahl pathway, as proposed by Valentine et al. (2004) [30]. The transfer of methyl groups by methyl-transferase to coenzyme-M may represent the bottleneck in the entire pathway, only becoming reversible under strong substrate limitation.

While differential reversibility could explain isotope fractionation for hydrogenotrophic methanogens, fractionation during acetoclastic methanogenesis (Figure 1A) is very small and not anywhere near isotopic equilibrium. Isotopic fractionation during acetoclastic methanogenesis could also not explain the large difference in δ^{13}C between CH$_4$ and CO$_2$ in marine sediments, while the intra-molecular difference between the methyl and carboxyl carbons in acetate is only in the range of 7‰–14‰ (Blair and Carter, 1992; Sugimoto and Wada, 1993) [35,56]. While acetate is split into a methyl-group and a carboxyl-group within the ACS-CODH cluster, a direct exchange of the methyl-C and the carboxyl-C would not be possible, due to the high energy barrier involved in cleaving the C=O double bond. Also, in most cases, acetate represents a small intermediate pool and cannot significantly fractionate against CH$_4$ and CO$_2$. Fractionation in the anabolic branch (Figure 2) is also not likely to be a major cause of ^{13}C-depleted methane, since archaeal lipids are themselves depleted in ^{13}C (e.g., Hinrichs et al., 2000; Pancost et al., 2000; Contreras et al., 2013) [57–59]. Isotope exchange may still occur through the MTP branch in facultative hydrogenotrophic *Methanosarcina* (capable of hydrogenotrophic and acetoclastic methanogensis). Indeed, *Methanosarcina* growing acetoclastically shows 20‰–30‰ fractionation, while fractionation in the obligate acetoclastic *Methanosaeta*, which is a common organism in marine sediments (Carr et al., 2017) [60], is entirely insignificant (Figure 1A).

Overall, isotopic equilibration through the WL pathway may represent the most predominant carbon-isotope fractionation mechanism in natural marine sediments. However, this concept requires further investigation. Radiotracer experiments with methanogens or incubations with purified enzymes could clarify some of the fundamental fractionation mechanisms.

2.4. The Effects of Substrate and Carbon Limitation

Energy limitation: It is well known that microorganisms in the deep biosphere often operate under extreme energy limitation (Jørgensen and D'Hondt, 2006) [37]. Under such conditions, enzymatic reactions are likely reversible, leading to isotopic exchange. An isotopic equilibration is possible as long as the reaction is also approaching a chemical equilibrium (Urey and Greiff, 1935) [40]. This is clearly the case for hydrogenotrophic methanogens under H$_2$ limitation, while ample inorganic carbon may still be available. This is also the case during AOM because the energy yield of AOM is minimal and subject to concentration levels of methane and sulphate. Yoshinaga et al. (2014) [39] observed an increasing reverse AOM flux and accordingly, more complete isotopic equilibration with decreasing sulphate concentration. In their model, equilibrium fractionation results from the difference in fractionation of the forward and reverse reactions within a diffusion-limited system.

Organic carbon substrate limitation: Also, organic carbon substrates are often limiting under deep-biosphere conditions. In particular, small molecules, such as acetate or methyl groups, are common intermediates produced by other, fermenting organisms at rather low rates. Under sluggish turnover rates, not only is the energy source limited, but also the carbon source, in which case, complete turnover may result in a minimal fractionation effect.

Inorganic carbon limitation: It is very unlikely that in normal marine sediments dissolved inorganic carbon (DIC) becomes limiting. This is because, according to Equation (4), for each mole of H_2 produced by fermentation of organic matter, approximately half a mole of CO_2 is produced (i.e., two moles of CO_2 per mole of CH_4). CO_2-limitation may, however, occur if an excess of H_2 is supplied from abiotic sources. Such observations have been made in off-ridge hydrothermal systems, such as the Lost-City hydrothermal field in the Atlantic, where large amounts of H_2 are produced due to alteration of ultramafic rocks by seawater (serpentinization; see McCollom and Bach, 2009) [61]. Ultramafic fluids are often highly alkaline and conducive to intense carbonate precipitation if they come in contact with seawater, i.e. their DIC content is usually very low. Under high H_2 content and concomitantly low DIC content, the system may indeed become DIC-limited. Under CO_2-limitation a complete Rayleigh effect would occur and, hence, isotope fractionation would become small. In these cases, biogenic methane production would produce CH_4 with a rather high ^{13}C content, thus mimicking an isotopic signature of methane that was generally thought to be of abiotic origin (Meister et al., 2018) [62]. The experiments by Miller et al. (2018) [33] seem to confirm that methanogenesis under alkaline conditions with strong Ca-carbonate supersaturation indeed produce methane with very small fractionation relative to CO_2.

Considering all these different effects, we may conclude that carbon-isotope equilibrium fractionation is most likely to prevail in normal marine sediments.

3. Carbon-Isotope Distribution in Porewater Systems

3.1. Isotope Effects in Diffusive Mixing Profiles

While the exact mechanisms of fractionation remain incompletely understood, and further experiments are necessary to shed light on this matter, a further aspect that needs consideration is how the fractionation manifests in isotopic differences in natural methane and DIC profiles in sedimentary porewater. In unlithified marine sediments, porosity is usually large, allowing for diffusive transport of solutes along concentration gradients. While advective transport may substantially contribute to methane and DIC transport (see Section 3.4), diffusive mixing alone leads to considerable complexity in isotopic distributions. The main C-constituents in sedimentary porewaters are DIC and methane. Although dissolved organic carbon (DOC) compounds can be detected, which are largely fermentation products, such as acetate, serving as intermediates in microbial metabolic networks, their concentration gradients are generally too small to give rise to significant diffusive transport. Even though their turnover may exhibit substantial fractionation effects (e.g., Heuer et al., 2008 [7]; Ijiri et al., 2012 [63]), these effects do not manifest in the end products if these compounds undergo quantitative production and consumption, resulting in a complete Rayleigh effect.

Diffusive mixing between seawater and porewater: Inorganic carbon dissolved in seawater has a $\delta^{13}C$ around 0‰, but its concentration is only on the order of 2 mmol/L. DIC concentration often steeply increases with depth in the sediment, often reaching 10 mmol/L within a few centimetres. Most DIC in the sediment is derived from dissimilatory degradation of organic carbon, exhibiting an isotopic composition on the order of −20‰, while the influence of the small amounts of DIC from seawater rapidly decreases with depth (Figure 4A). Diffusive mixing is then manifested as a mixing hyperbola, resulting in rather negative values at shallow depth in a diffusive boundary layer (Zeebe, 2007) [64].

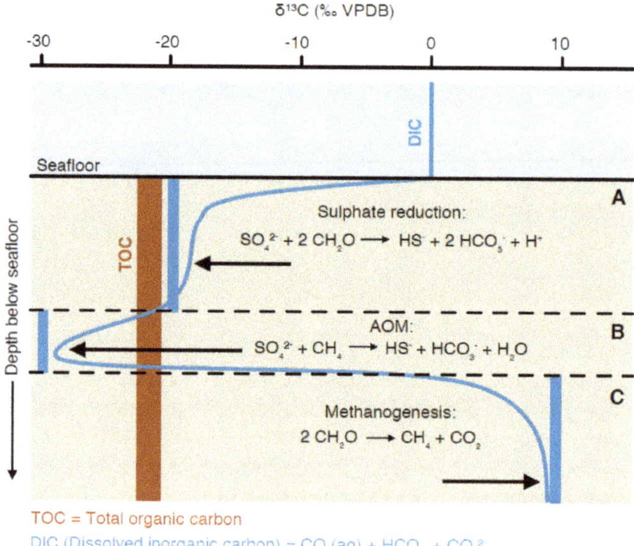

Figure 4. Scheme of isotope fractionation effects due to metabolic activity on δ^{13}C of DIC across the SMT. Essentially, three different zones can be distinguished based on metabolic activity and isotopic signatures in the porewater (from top to bottom): (**A**) Sulphate reduction zone; (**B**) SMT showing anaerobic oxidation of methane; (**C**) methanogenic zone. The thick blue bars represent the isotopic compositions of the instantaneously produced metabolic DIC; the thin blue line represents the resulting isotope profile of DIC.

Isotope profiles at the SMT: Further complexity arises near the SMT, where even more ^{13}C-depleted DIC from AOM is mixing with DIC from the sulphate reduction zone (Figure 4B). Because isotopically light methane is quantitatively converted to DIC, δ^{13}C of DIC may become much more negative than δ^{13}C of TOC at the SMTZ. Extremely negative δ^{13}C$_{\text{DIC}}$ values (<−90‰) were measured in deep continental crust fracture systems (Drake et al., 2015, 2017 [65,66]). However, such negative values are only possible under DIC limitation, which is not usually the case in marine sediments (see above section). The negative effect from AOM is usually balanced by diffusive mixing with ^{13}C-enriched DIC from the methanogenic zone (Figure 4C).

Isotope profiles in the methanogenic zone: As shown above and in Figure 1A, fractionation effects are rather strong during hydrogenotrophic methanogenesis, resulting in isotopically light CH$_4$ (δ^{13}C < −60‰) and isotopically heavy DIC (δ^{13}C > 0‰; Figure 4C), whereby some intermediate CO$_2$ is also produced by fermentation (Equation (4)). Acetoclastic methanogenesis may occur, as indicated by enriched δ^{13}C values in acetate observed by Heuer et al. (2008) [7]. However, this process probably does not occur alone and does not explain the large isotopic differences between DIC and CH$_4$ observed in most methanogenic zones. As a result of diffusive mixing, DIC shows a mixing curve between the negative values at the SMT and moderately positive values in the methanogenic zone (Figure 4C).

3.2. Towards a Simulation of Carbon-Isotope Profiles

Closed-system Rayleigh models: Several studies have been dedicated to understanding these effects using closed-system Rayleigh fractionation models (Nissenbaum et al., 1972 [67]; Claypool and Kaplan, 1974 [6]; Whiticar and Faber, 1986 [19]; and Paull et al., 2000 [68]). δ^{13}C was calculated as a function of the fraction of CO$_2$ converted to CH$_4$ (Rayleigh function), without CO$_2$ production by fermentation, but including the input by CO$_2$ advection and the removal by carbonate precipitation. The model was solved for separate compartments in a sedimentary porewater profile, yet, the model does not simulate diffusive mixing. It would be incorrect to fit a Rayleigh curve, representing an exponential function, to a porewater profile showing a mixing hyperbola.

Reaction-diffusion models: Instead, Alperin et al. (1988) [20] was among the first to use a numerical reaction-transport model to simulate diffusion and fractionation of CH_4 above the SMT. By fixing the $\delta^{13}C_{CH4}$ at the SMT and near the sediment surface as boundary conditions, he realized that the shape of the curve in the sulphate-reduction zone is very sensitive to the kinetic fractionation factor of AOM (the forward reaction). Chatterjee et al. (2011) [69] simulated $\delta^{13}C_{DIC}$ through the SMT using a reaction-transport model, which very well reproduced the measured profiles from several sites at Hydrate Ridge and in the Gulf of Mexico.

Models including isotopic equilibration: Despite the above studies, it still remained unclear what caused the strongly negative $\delta^{13}C_{CH4}$ values at the SMT. The key insight came from Yoshinaga et al. (2014) [39], demonstrating a reverse flux and isotope equilibration during AOM. These authors could satisfactorily reproduce $\delta^{13}C_{CH4}$ profiles across the SMT at the Cascadia Margin (Figure 5; Bull's eye vent; Pohlman et al., 2008) [70], and this concept was successfully applied to reproduce profiles in the South China Sea (Wu et al., 2018 [71]; Chuang et al., 2019 [72]). Now it is possible to simulate both $\delta^{13}C_{CH4}$ and $\delta^{13}C_{DIC}$ including the complete stoichiometries of metabolic reactions in diffusive sedimentary systems. Meister et al. (2019b) [55] developed and tested such a model system, including the SMT and methanogenic zone over a depth range of 200 m. The modelling confirmed previous insights that isotopic equilibration at the SMT must occur.

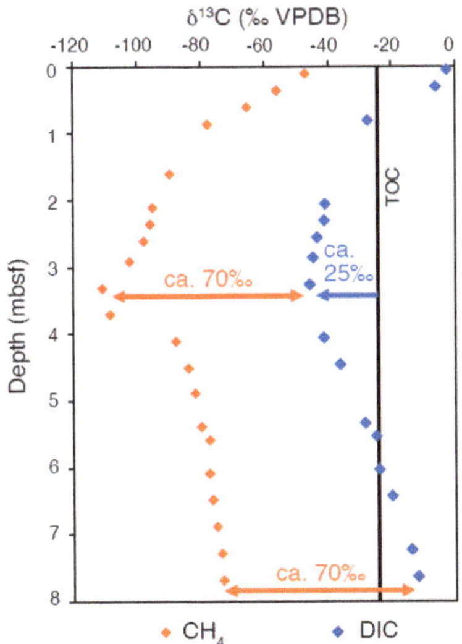

Figure 5. Measured $\delta^{13}C_{CH4}$ and $\delta^{13}C_{DIC}$ profiles from SMTs at Site C-2, Bullseye vent, Cascadia Margin (data from Pohlman et al., 2008 [56]; figure modified from Meister et al., 2019b [35]).

3.3. Assessing the Factors Influencing Carbon-Isotope Profiles

Having a full reaction-diffusion model at hand, Meister et al. (2019b) [55] discuss the sensitivity of carbon-isotope profiles to variations in organic matter burial and fractionation factors through the SMT and into the methanogenic zone. One observation is that fractionation factors determined in culture experiments with *Methanosarcina barkeri* under different conditions (Krzycki et al., 1987 [26]; Gelwicks et al., 1994 [27]; Londry, et al., 2008 [31]; etc.; Figure 1) result in too small isotopic differences between CH_4 and CO_2 compared to the differences observed in natural environments. Larger fractionation was observed in cultures using strains of *Methanococcus*, an obligate hydrogenotrophic methanogen, under substrate limitation (Botz et al., 1996) [28] (Figure 1). These larger fractionation factors would be necessary to reproduce the measured porewater profile (Meister et al., 2019b) [55], including the effect of CO_2 production by fermentation as shown by Equation (4).

Overall, methanogenesis results in ^{13}C-depleted CH_4 and ^{13}C-enriched DIC, and, from a stoichiometric point of view, isotopic values of CH_4 and DIC should be symmetrical with respect to the organic matter substrate from which all DIC and CH_4 originates, independent of the pathway. The $\delta^{13}C_{CH4}$ and $\delta^{13}C_{DIC}$ approach a symmetrical distribution relative to $\delta^{13}C_{TOC}$, which is indeed observed in many measured isotope profiles (Figure 6A). While the (apparent) separation factor ε between $\delta^{13}C_{CH4}$ and $\delta^{13}C_{DIC}$ remains constant, the $\delta^{13}C_{CH4}$–$\delta^{13}C_{DIC}$ couple may be shifted to lower values if the organic matter decay is slow or if most organic matter rapidly decays in the sulphate reduction zone (resulting in a deep SMT and curved sulphate profile; Meister et al., 2013a) [73]. This is because the $\delta^{13}C_{DIC}$ would then be more influenced by ^{13}C-depleted DIC diffusing down from the sulphate-reduction zone (Figure 2B). However, no shift to more positive values (as seen in Figure 6B; Heuer et al., 2009) [7] can ever occur in a steady-state diffusive system. In general, it is difficult to reproduce $\delta^{13}C_{DIC}$ more positive than about 10‰ and impossible to produce the 35‰ observed by Heuer et al. (2009) [7].

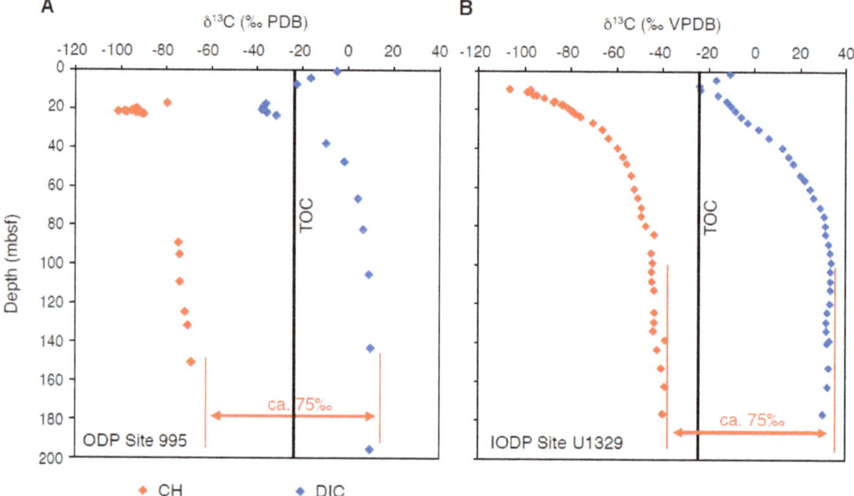

Figure 6. Measured $\delta^{13}C_{CH4}$ and $\delta^{13}C_{DIC}$ profiles from natural methanogenic zones: (**A**) Blake Ridge, ODP Site 995 (Paull et al., 2000) [54]: The profiles largely approach symmetry with respect to $\delta^{13}C_{TOC}$ in the methanogenic zone. (**B**) Cascadia Margin, IODP Site U1329 (Heuer et al., 2009) [7]: Profiles are shifted towards more positive values, with DIC reaching +35‰ at 100 mbsf.

In conclusion, most features of δ^{13}C profiles of CH_4 and DIC in marine porewaters can be well reproduced with numerical reaction-diffusion models, but to explain very positive values in DIC observed in some methanogenic zones further advective or, perhaps, non-steady state conditions need to be taken into account.

3.4. Non-Steady-State Effects and Gas Transport

Non-steady state effects: Although it is clear that porewater profiles are often not in a steady state due to changes in sediment deposition and microbial activity (e.g., Dale et al., 2008a; Contreras et al., 2013) [59,74], the effects on isotope profiles are still poorly assessed. It is possible that upward and downward migration of the SMT results in a temporary offset of isotope profiles with respect to the redox zonation. Also, increase and decrease in overall microbial activity may shift the isotopic signatures (Meister et al., 2019b) [55], as will be further discussed below.

Methane and CO_2 rise: In organic carbon-rich sediments both CH_4- and CO_2-gas fugacities may exceed solubility in porewater, which then leads to exsolution of a gas phase. Gas systematics are rather complex, as many factors play a role, while gas transport is often unpredictable, abrupt, and anisotropic, i.e., it follows discrete conduits. Starting from oversaturated CH_4, rates of gas bubble formation can be calculated according to Mogollón et al. (2009) [75]. While methane gas saturation is soon reached, CO_2 may not readily be oversaturated due to the buffering effect of porewater alkalinity (see discussion below). Nevertheless, once a gas phase is present, as methane bubbles, CO_2 readily equilibrates with the gas phase, following Le Chatelier's principle (Smrzka et al., 2015) [76]. While parts of the methane bubbles adhere to the sediment particles by capillary forces, the rise velocity of methane bubbles can be calculated using the formulations of Boudreau (2012) [77]. Thereby, it is critical to consider that methane bubble rise is facilitated if conduits, such as fractures or porous sediment, are present. Once the bubbles reach a zone with a lower CH_4 fugacity near the SMT, bubbles start to re-dissolve into the aqueous phase, but depending on the dissolution kinetics (see Mogollón et al., 2009) [75] and the thickness of the sulphate zone, parts of the methane bubbles may reach the seafloor and escape to the water column (seepage; e.g., Dale et al., 2008b) [78]. While methane seepage out of marine sediments is not commonly reported to be CO_2-rich, parts of the CO_2 are probably re-dissolved due to the pH-buffering effect of AOM at the SMT. The diffusion constant of CO_2 only shows minor dependence on isotopic mass (Zeebe et al., 2011) [79]; however, the fact that CO_2, which is depleted in ^{13}C by ca. 9‰ relative to HCO_3^-, rises more rapidly (due to a higher diffusion constant, or in the gas phase) may also have an effect on the overall isotopic composition of DIC. However, this effect is limited by the carbonate equilibrium, preventing a runaway Rayleigh effect.

Deep methane sources: Methane gas may not only evolve within the considered sediment sequence, but may rise from deeper intervals of organic carbon-rich sediments, from the gas hydrate stability zone (e.g., Borowski et al., 1999; Kennett et al., 2000) [80,81], or even from the thermogenic zone, where methane is generated from the thermogenic decomposition of organic matter. In particular, thermogenic methane is generally less depleted in ^{13}C (e.g., Whiticar, 1999) [82]. It still remains unexplained how DIC with extremely positive $\delta^{13}C_{DIC}$ values could be produced in the methanogenic zone. However, if thermogenic methane could re-equilibrate, isotopically, with the DIC pool through the WL-pathway, this could explain the positive values observed at the Cascadia margin (Heuer et al., 2009) [7], where the CH_4-DIC couple is shifted to more positive $\delta^{13}C$ than symmetry with respect to TOC (Figure 6B). Testing such scenarios requires more elaborate numerical models under non-steady-state conditions.

4. The Isotope Signature of Diagenetic Carbonates

4.1. Processes Inducing Carbonate Formation in the Deep Biosphere

Before discussing how carbon-isotope patterns characteristic for particular biogeochemical zones are preserved in diagenetic carbonates, the factors controlling authigenic (incl. diagenetic) carbonate formation in sediments need to be briefly summarized. Generally, carbonates precipitate due to an increase of the saturation state (here expressed as saturation index $SI = \log IAP - \log K_{SP}$, where IAP is the ion activity product and K_{SP} is the solubility product). Most commonly, cations (mainly Ca^{2+} and Mg^{2+}) are sufficiently supplied from seawater in the uppermost few metres (Baker and Burns, 1985; Meister et al., 2007) [11,83] or sometimes through deep circulating fluids (Meister et al., 2011) [84]. Carbonate saturation can be significantly increased due to microbial metabolic activity, whereby it is still debated which microbial processes indeed can induce carbonate precipitation. Under marine conditions, sulphate reduction, which is producing DIC and alkalinity at a 1:1 ratio, may even lead to a lowering of the saturation state due to a drop in pH (Meister, 2013 [85]; and references therein), unless most of the 28 mmol/L of sulphate in seawater are turned over. In contrast, AOM produces two moles of alkalinity per mole of DIC and, thus, efficiently increases the SI of carbonates (Moore et al.,

2004; Ussler III. et al., 2008; Meister, 2013) [85–87]. Furthermore, methanogenesis always produces CO_2 and no alkalinity (as seen from Equation (3)), but the acidification effect is largely buffered by alkalinity produced near the SMT and the release of ammonia. Further alkalinity may originate from the alteration of silicates (mainly volcanic glass but possibly also by clay minerals; Wallmann et al., 2008; Meister et al., 2011; Wehrmann et al., 2016) [84,88,89]. These sources of alkalinity production, perhaps in combination with exsolution of CO_2 via methane bubbles, may prevent carbonate undersaturation in the methanogenic zone. Although these effects have not been precisely quantified yet, and it remains unclear how focused diagenetic beds of carbonate can form in the methanogenic zone, it is most likely due to the dynamics of a supersaturation front (cf. Moore et al., 2004) [86].

4.2. Controls of $\delta^{13}C$ Composition of Diagenetic Carbonates

Carbonate precipitation itself is subject to fractionation effects, whereby equilibrium fractionation prevails at slow precipitation rates observed in the deep biosphere (Turner, 1982) [90]. The carbonate mineral phase is, in most cases, enriched in ^{13}C by a few permil relative to the inorganic carbon (~2‰ for calcite; Deines et al., 1974) [91]. For dolomite, the separation factor relative to CO_2 is on the order of 12‰–14‰ at ambient temperatures (Ohmoto and Rye, 1979; Golyshev et al., 1981) [92,93]. Subtracting the isotope effect of 9‰ between CO_2 and HCO_3^- (Mook, 1974) [94] results in a range of 3‰–5‰ for dolomite and HCO_3^-. As a result of this fractionation effect, it was suggested that the residual DIC is depleted in ^{13}C due to a Rayleigh effect (Michaelis et al., 1985) [95]. However, since carbonate precipitation is limited by the production of alkalinity and the supply of major cations from seawater, while DIC is usually not limited in the deep biosphere, as it is produced in ample amounts from microbial dissimilation reactions, carbonate precipitation has most likely a minor effect on the isotopic composition of DIC. This has also been confirmed by model calculations (e.g., Chuang et al., 2019) [72]. Thus, the carbon-isotope signature of the porewater becomes trapped in the diagenetic carbonate, providing a signature for past biogeochemical conditions at the location and time of precipitation.

Suboxic vs. anoxic zones: It is often seen in carbonates, especially if they occur in organic carbon-rich sediments, that $\delta^{13}C$ values are in a range between 0‰ and −10‰, but not as negative as to indicate a signature typical for a sulphate reduction zone. This could be the result of a precipitation in the top few centimetres below the sediment surface, where carbon isotopes follow a mixing hyperbola. Typically, in suboxic sediments, where the dissimilatory rates are moderate, and the redox zonation accordingly expanded, $\delta^{13}C$ values fall into this intermediate range, as observed in Ca-rich rhodochrosite occurring within mottled and bioturbated sediments of the Eastern Equatorial Pacific (Meister et al., 2009) [96]. Thereby Fe- and Mn-reduction may contribute to carbonate supersaturation (Kasina et al., 2017, and references therein) [97]. Intermediate values may also occur in shallow sulphate-reduction zones, e.g., in bituminous sediments, where laminae of authigenic carbonate form just below the sediment/water interface and, hence, early with respect to burial along the mixing gradient (cf. the Triassic Besano Fm., Ticino, Switzerland; Bernasconi et al., 1994 [98]; see discussion in Meister et al., 2013b) [99]. Alternatively, the isotope values may represent a mixture of different carbonate phases of different origin, e.g., dolomite mud from an adjacent platform, showing normal marine isotope values. Therefore, a further petrographic analysis is often necessary to determine the origin of the carbonate, in order to interpret its carbon-isotope signature.

Sulphate-methane transition zone: While sulphate reduction alone rather lowers the saturation state of carbonates, an early onset of AOM has been suggested to induce the formation of carbonates, such as carbonate concretions in organic carbon-rich shale of the Santana Fm. (Brazil; Heimhofer et al., 2018) [100] at very shallow depths. Shallow SMT zones are well documented from modern settings (e.g., Thang et al., 2013) [101] and they may indeed induce carbonate cementation (e.g., Jørgensen et al., 1992) [102]. However, an actual AOM signature in $\delta^{13}C$ with values below −35‰ is only exceptionally preserved, such as in a dolomite layer at the Peru Trench at 6.5 m below seafloor (mbsf; Meister et al., 2007) [11]. Instead, a great range of carbon-isotope values have been reported from dolomite layers intercalated in organic carbon-rich diatom ooze drilled from upwelling regions offshore California

(Pisciotto and Mahoney, 1981) [103], in the Gulf of California (Kelts and McKenzie, 1982) [104], or in Miocene diatomite of the Monterey Fm. (California; Murata et al., 1969; Kelts and McKenzie; 1984) [9,105]. Also, Rodriguez et al. (2000) [106] report strongly positive δ^{13}C values in siderites from a methanogenic zone at Blake Ridge. While the positive δ^{13}C values were interpreted as a result of precipitation in the methanogenic zone, negative values in dolomites of the Messinan Tripoli Fm. in Sicily were interpreted as indicative of precipitation in the sulphate-reduction zone. This explanation seems obvious, but it still remains unclear what caused precipitation of carbonates in the methanogenic zone, as methanogenesis should not per se lead to a focused supersaturation of carbonates.

Deep methanogenic zone: A case in which dolomite cements are indeed observed to form in the methanogenic zone is OPD Site 1230, located in the Peru Trench. This site, at a water depth of 5000 m, is located on the Peruvian accretionary prism, where the sedimentary succession is dissected by a fault zone at 230 mbsf). A dolomite breccia was drilled at this depth, showing more radiogenic ^{87}Sr/^{86}Sr ratios than modern seawater, indicating precipitation from a fluid that was derived from interaction with continental basement rocks, deep in the prism (Meister et al., 2011) [84], presumably delivering alkalinity and Ca^{2+} to induce dolomite precipitation. While this site represents a special case, it is still not understandable how dolomite can otherwise form in a methanogenic zone.

4.3. Interpreting δ^{13}C Archives Through Time

Carbon isotopes were measured in diagenetic dolomites through a 150 m thick interval on the Peru Margin, showing varying δ^{13}C (Figure 7A–C; ODP Site 1229; Meister et al., 2007) [11]. The dolomites also show ^{87}Sr/^{86}Sr ratios near to Pleistocene–Holocene seawater, while the ratios in the porewater strongly decrease with depth, indicating that the dolomites formed near to the sediment–water interface. The dolomites thus formed in the past and document an actively evolving biosphere through time.

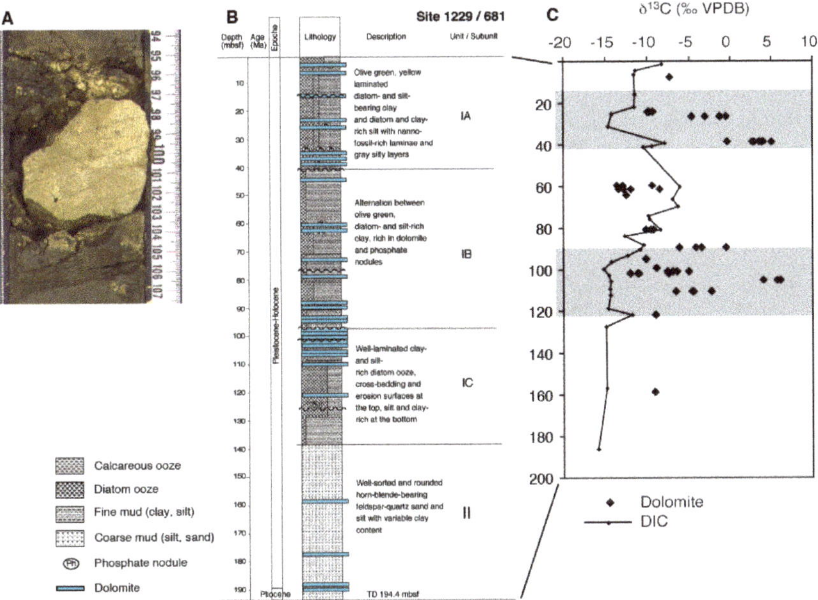

Figure 7. Patterns of carbon isotopes preserved in diagenetic carbonates from the Peru Margin: (**A**) Fragment of discrete, hard lithified, diagenetic dolomite; (**B**) distribution of dolomite layers through the sequence at ODP Site 1229; (**C**) carbon-isotope values in diagenetic dolomites in comparison to carbon-isotope composition of the present porewater.

Episodic carbonate formation: The dolomite beds do not present uniform conditions through time, as otherwise fine-grained dolomite would have been homogeneously distributed throughout the sediment. Instead precipitation must have occurred episodically, at focused locations. Based on their regular spacing on the order of glacial–interglacial cycles in the sediment, Compton (1988) [107] proposed that diagenetic dolomite beds in the Monterey Fm. could be linked to Milankovitch cyclicity in sediment deposition. Meister et al. (2008) [108] followed this idea by showing that oxygen isotopes in dolomites, that occur with a spacing of ca. 10 m, reflect marine $\delta^{18}O$ values and bottom water temperatures of glacial periods on the Peru Margin. Contreras et al. (2013) [59] found enrichments of dolomite, barite, and isotopically light archaeol ($\delta^{13}C = -73‰$) as an imprint of an earlier shallow SMT zone far above the present SMT, which has shifted downward since then. In this manner, the isotopic signature of archaeol can be explained by fractionation as part of the Wood–Ljungdahl pathway (see discussion above). Reaction-transport modelling then confirmed that an upward and downward migration over 30 meters within the time frame of 100 ka is feasible. In conclusion, the dolomite layers on the Peru Margin formed in the aftermath of a rapid deposition of an organic carbon-rich, interglacial sediment layer that, during its burial, triggered a temporary onset of a shallow SMT zone. During this time, a dolomite layer formed at the upper SMT.

Dynamics in carbon-isotope preservation: While the episodic precipitation of dolomites can be explained by a 100 ka cyclicity in deep biosphere activity, longer-term changes on the order of several 100 ka are superimposed and manifested in the $\delta^{13}C$ record. In theory, three explanations can be proposed (Figure 8A–C): (A) Precipitation in different zones as suggested by Kelts and McKenzie (1984) [9], whereby the carbon-isotope signature is controlled by the depth of a carbonate saturation front relative to the redox zonation and $\delta^{13}C$ profile in the porewater. The saturation front could be uncoupled from the redox zonation due to outgassing of CO_2 from the methanogenic zone. (B) At a shallow SMT, outgassing of CH_4 is frequently observed (e.g., Dale et al., 2008b) [78], resulting in the loss of ^{13}C-depleted carbon and accordingly, less negative $\delta^{13}C_{DIC}$ near the SMT. This mechanism can be well reproduced with reaction transport modelling (Meister et al., 2019b, Figure 7 therein) [55]. (C) Due to an increase in methanogenic activity, $\delta^{13}C$ in both CH_4 and DIC may increase in the methanogenic zone. This is clearly the case at Peru Margin ODP Site 1229, where the modern-day $\delta^{13}C_{DIC}$ is negative in the methanogenic zone, but was positive in the past (Meister et al., 2019a; and references therein) [13]. The observation that the variations of $\delta^{13}C$ in diagenetic dolomites are coupled to variations of $\delta^{34}S$ in pyrite provides independent evidence that two episodes of enhanced deep-biosphere activity involving stronger methanogenic activity occurred throughout the Pleistocene. Most likely, a combination of the effects A-C occurs in a dynamic way. Ultimately, reaction-transport modelling under non-steady-state conditions will clarify the mechanisms that generated these diagenetic records.

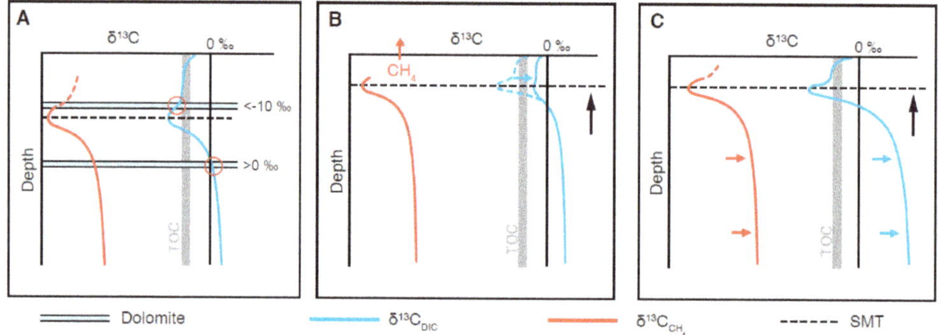

Figure 8. Possible scenarios of carbon-isotope incorporation in diagenetic dolomites: (**A**) Dolomite precipitation in different redox zones; (**B**) dolomite formation at the SMT showing variable $\delta^{13}C_{DIC}$ as a result of CH_4 escape; (**C**) changing $\delta^{13}C_{DIC}$ due to variations in methanogenic activity.

5. Implications for the Global Carbon Cycle

5.1. The Importance of Diagenetic Carbonates for Global Carbon Fluxes

Carbonates are besides sedimentary organic carbon the largest sink of CO_2 from the exogenic cycle. The capacity of the ocean to precipitate carbonate relies on the delivery of alkalinity from continental silicate weathering. Also, a significant portion of carbonate forms as a result of submarine alteration of ultramafic rocks, so that subduction of ophicarbonates substantially contributes to the global carbon cycle (e.g., Alt and Teagle, 1999) [109]. Diagenetic carbonates forming as a result of dissimilatory microbial activity in marine sediments provide a further carbon sink. Their formation does not rely on alkalinity supplied by continental silicate weathering; they are induced by anaerobic terminal electron accepting processes acting as an alkalinity pump. This alkalinity remains in the sediment, as long as sulphide is trapped as iron sulphides and not re-oxidized at the sediment surface. Microbial alkalinity production contributes to retaining a significant portion of DIC, derived from organic matter, that otherwise would be cycled back to the water column and atmosphere (Figure 9). Schrag et al. (2013) [8] estimated that diagenetic carbonates significantly contribute to the global carbonate burial flux, in particular at times of widespread anoxia, and that this flux also may substantially affect global carbon-isotope composition. Burial of isotopically light carbon, as diagenetic carbonate, would contribute to an increase in $\delta^{13}C$ in ocean and atmosphere. In turn, formation of isotopically heavy diagenetic carbonates would decrease $\delta^{13}C$ in the ocean and atmosphere, and therefore, understanding the controls of methanogenic carbonate formation would be significant for assessing the deep biosphere's influence on the carbon cycle.

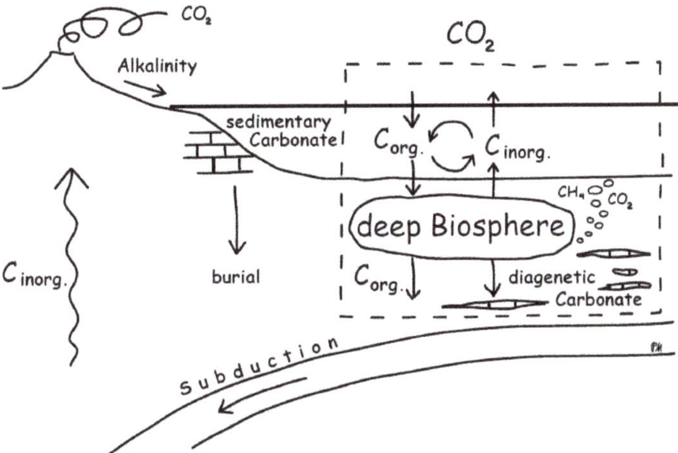

Figure 9. Schematic drawing representing the role of the deep biosphere in the global carbon cycle. Parts of the organic carbon deposited on the seafloor are re-mineralized and cycled back to the water column. Some inorganic carbon is buried as diagenetic carbonate. The latter is mainly induced by additional alkalinity from anaerobic metabolisms and is decoupled from sedimentary carbonate induced by alkalinity from continental weathering. Most carbon is stored in the rock record, but the deep biosphere has a significant effect on what goes into this reservoir.

5.2. Diagenetic vs. Atmospheric Signatures?

Certain time periods in Earth's history stand out due to significant excursions in the global carbon-isotope composition. These excursions may be largely influenced by the activity of the deep biosphere. For example, methane released from the dissociation of gas hydrates as a result of warming of bottom water during the Palaeocene–Eocene thermal maximum has caused a perturbation in the carbon cycle, manifested in a negative excursion in the atmospheric $\delta^{13}C$ (Dickens, 1997) [110]. Also, positive excursions occurred during the Neoproterozoic (Knoll et al., 1986) [111] or near the Permian–Triassic boundary (Payne et al., 2004) [112], presumably due to enhanced burial of organic carbon.

While signatures in the carbon-isotope record can often be globally correlated, it still has to be confirmed that these are not results of diagenetic overprint. One indication that the signatures indeed represent the conditions in the water column would be that isotope records in the organic carbon show a similar excursion as the inorganic carbon, i.e., the organic and inorganic fractions run in parallel, offset by 20‰–30‰, depending on the separation factor of the primary production (Knoll et al., 1986) [111]. Isotopically light signatures in organic carbon from the Peru Margin were initially interpreted as signatures of a large-scale methane escape to the water column along the Peru Margin, presumably as a result of gas hydrate dissociation (Wefer et al., 1994) [113]. However, the finding of accumulated isotopically light archaeol, in combination with barite and dolomite enrichments, supports the concept that these signatures instead represent the imprint of a former SMT (Contreras et al., 2013) [59] and are thus a diagenetic feature. Similarly, Louis-Schmid et al. (2007) [114] found that a 4‰ negative excursion in a Late Jurassic hemipelagic succession at Beauvoisin (SE France) is the result of local anaerobic methane oxidation and precipitation of authigenic carbonates, rather than a global signal.

An even more extreme carbon-isotope excursion can be found in the Palaeoproterozoic record. The so-called Lomagundi–Jatuli event (ca. 2.2–2.1 Ga) lasted more than 100 Ma and is the most positive excursion in $\delta^{13}C$ (up to +14‰) in Earth's history. It was thought that this excursion is the result of strongly enhanced organic carbon burial rates as a result of the onset of oxygenic photosynthesis at the time of the great oxidation event, i.e., when Earth's atmosphere first became oxidized (Schidlowski et al., 1984) [115]. However, the excursion is not observed in the organic carbon record, which casts some doubt on the explanation that organic carbon burial is responsible for such a large isotope excursion. Hayes and Waldbauer (2006) [116] challenged this interpretation, using a global model of carbon and energy fluxes. According to their calculation, such a large carbon burial would have resulted in unrealistically high oxygen concentrations (up to 100 times the present level). Instead, they suggested that organic carbon production was enhanced and, while the sulphur cycle was not fully developed yet, organic matter would have been converted to CH_4 and CO_2 at shallow depths, with minimal AOM taking place. Escape of isotopically light methane would have led to an isotopically heavier residual DIC pool. Presumably, CO_2 outgassing was partially prevented by a high alkalinity of the ocean due to silicate weathering, causing carbonates to precipitate. In other words, the signature observed in the carbonate record represents a diagenetic signal and, since this signal occurs simultaneously around the Earth, it would represent a "global diagenetic event".

That this process indeed can occur has been demonstrated for modern stromatolites in a brackish to seasonally evaporative, coastal pond in Brazil (Lagoa Salgada), showing extremely positive carbon-isotope signatures of >15‰ (Birgel et al., 2015) [117]. In fact, this system has been suggested as a modern analogue to explain the large excursions in $\delta^{13}C$ in the aftermath of the GOE. Also, in this case, carbon-isotope signatures provide an archive of major dynamics in the "shallow" deep biosphere of global scale.

6. Conclusions

This synopsis provides an overview of processes affecting carbon isotopes preserved in diagenetic carbonates and the potential of these isotopic signatures to serve as indicators of past conditions in the marine deep biosphere. Culture experiments using different organisms under a range of different growth conditions, as well as radiocarbon tracer experiments, provide insight on potential fractionation mechanisms as part of the Wood–Ljungdahl pathway, which is the main methane metabolizing pathway. It is indicated that during AOM, but also during methanogenesis, isotopic equilibration between CO_2 and CH_4 could play a role and could explain natural isotope distributions, which are near to a distribution expected at thermodynamic equilibrium. Taking these effects into account, isotopic distributions in diffusive porewater profiles, in response to organic matter content, reactivity, and sedimentation rate, are largely reproducible using numerical reaction-transport models.

Carbon-isotope signatures preserved in diagenetic carbonates show strongly variable values, but they are still not fully understood. Most likely, diagenetic carbonates form episodically, reflecting dynamic conditions and probably upward and downward shifts in the redox zonation. Most likely, activity of the deep biosphere is linked to cyclicity in oceanographic conditions and sediment deposition on different timescales, partially affecting the isotopic signatures. Advective transport of gas phase CH_4 and CO_2 may also affect the carbonate isotope profiles. Carbonate precipitation is generally induced by alkalinity production by microbial activity, most importantly, AOM and ammonium release, but perhaps also silicate alteration. Upon discharge of CO_2 from the methanogenic zone, the supersaturation front may be decoupled from the depth of the SMT. Most urgently, though, numerical models under non-steady-state conditions are needed to disentangle such dynamics and to resolve peculiar carbon-isotope distributions in the deep-time geological record. This will ultimately reveal the potential of carbon-isotope signatures in diagenetic carbonates as archives of past deep-biosphere conditions.

Author Contributions: P.M. contributed the geochemical parts. C.R. contributed the biochemistry parts.

Funding: The work was partially supported by the through the Marie Skłodowska-Curie Actions Research Fellowship Programme: Project TRIADOL (grant agreement no. 626025) and the Department of Geodynamics and Sedimentology at the University of Vienna. Open Access Funding by the University of Vienna.

Acknowledgments: We thank S. Rittmann for comments related to the Wood-Ljungdahl pathway, as well as M.E. Böttcher, B. Liu, and B.B. Jørgensen for insightful discussions. We also thank three anonymous reviewers for constructive comments to the manuscript.

Conflicts of Interest: The authors declare no conflict of interest.

References

1. Broeker, W.S. A boundary condition on the evolution of atmospheric oxygen. *J. Geophys. Res.* **1970**, *75*, 3553–3557. [CrossRef]
2. Edwards, K.J.; Becker, K.; Colwell, F. The deep, dark energy biosphere: Intraterrestrial life on Earth. *Annu. Rev. Earth Planet. Sci.* **2012**, *40*, 551–568. [CrossRef]
3. Froelich, P.N.; Klinkhammer, G.P.; Bender, M.L.; Luedtke, N.A.; Heath, G.R.; Cullen, D.; Dauphin, P.; Hammond, D.; Hartman, B.; Maynard, V. Early oxidation of organic matter in pelagic sediments of the eastern equatorial Atlantic: Suboxic diagenesis. *Geochim. Cosmochim. Acta* **1979**, *43*, 1075–1090. [CrossRef]
4. D'Hondt, S.; Jørgensen, B.B.; Miller, J.; Batzke, A.; Blake, R.; Cragg, B.A.; Cypionka, H.; Dickens, G.R.; Ferdelman, T.; Hinrichs, K.U.; et al. Distributions of Microbial Activities in Deep Subseafloor Sediments. *Science* **2004**, *306*, 2216–2221. [CrossRef] [PubMed]
5. Hayes, J.M.; Popp, B.N.; Takigiku, R.; Johnson, M.W. An isotopic study of biogeochemical relationships between carbonates and organic carbon in the Greenhorn Formation. *Geochim. Cosmochim. Acta* **1989**, *53*, 2961–2972. [CrossRef]
6. Claypool, G.E.; Kaplan, I.R. The origin and distribution of methane in marine sediments. In *Natural Gases in Marine Sediments*; Kaplan, I.R., Ed.; Plenum Press: New York, NY, USA, 1974; pp. 99–140.
7. Heuer, V.B.; Pohlman, J.W.; Torres, M.E.; Elvert, M.; Hinrichs, K.U. The stable carbon isotope biogeochemistry of acetate and other dissolved carbon species in deep sub-seafloor sediments at the northern Cascadia Margin. *Geochim. Cosmochim. Acta* **2009**, *73*, 3323–3336. [CrossRef]
8. Schrag, D.P.; Higgins, J.A.; Macdonald, F.A.; Johnston, D.T. Authigenic carbonate and the history of the global carbon cycle. *Science* **2013**, *339*, 540–543. [CrossRef]
9. Kelts, K.; McKenzie, J.A. A comparison of anoxic dolomite from deep-sea sediments: Quaternary Gulf of California and Messinian Tripoli Formation of Sicily. In *Dolomites of the Monterey Formation and Other Organic-Rich Units*; Garrison, R.E., Kastner, M., Zenger, D.H., Eds.; Pacific Section SEPM: Los Angeles, CA, USA, 1984; Volume 41, pp. 19–28.
10. Malone, M.J.; Claypool, G.; Martin, J.B.; Dickens, G.R. Variable methane fluxes in shallow marine systems over geologic time: The composition and origin of pore waters and authigenic carbonates on the New Jersey shelf. *Mar. Geol.* **2002**, *189*, 175–196. [CrossRef]

11. Meister, P.; Bernasconi, S.; McKenzie, J.A.; Vasconcelos, C.; Frank, M.; Gutjahr, M.; Schrag, D. Dolomite formation in the dynamic deep biosphere: Results from the Peru Margin (ODP Leg 201). *Sedimentology* **2007**, *54*, 1007–1032. [CrossRef]

12. Meister, P. For the deep biosphere, the present is not always the key to the past: What we can learn from the geological record. *Terra Nova* **2015**, *27*, 400–408. [CrossRef]

13. Meister, P.; Brunner, B.; Picard, A.; Böttcher, M.E.; Jørgensen, B.B. Sulphur and carbon isotopes as tracers of past sub-seafloor microbial activity. *Nat. Sci. Rep.* **2019**, *9*, 604. [CrossRef] [PubMed]

14. Hoefs, J. *Stable Isotope Geochemistry*, 8th ed.; Springer: Berlin/Heidelberg, Germany, 2018; p. 437.

15. Alperin, M.J.; Blair, N.E.; Albert, D.B.; Hoehler, T.M.; Martens, C.S. Factors that control the stable carbon isotopic composition of methane produced in an anoxic marine sediment. *Glob. Biogeochem. Cycles* **1992**, *6*, 271–291. [CrossRef]

16. Hayes, J. Factors controlling ^{13}C contents of sedimentary organic compounds: Principles and evidence. *Mar. Geol.* **1993**, *113*, 111–125. [CrossRef]

17. Hayes, J.M. *An Introduction to Isotopic Calculations*; Woods Hole Oceanographic Institution: Woods Hole, MA, USA, 2004.

18. Londry, K.L.; Des Marais, D.J. Stable carbon isotope fractionation by sulfate-reducing bacteria. *Appl. Environ. Microbiol.* **2003**, *69*, 2942–2949. [CrossRef] [PubMed]

19. Whiticar, M.J.; Faber, E. Methane oxidation in sediment and water column environments—Isotope evidence. *Org. Geochem.* **1986**, *10*, 759–768. [CrossRef]

20. Alperin, M.J.; Reeburgh, W.S.; Whiticar, M.J. Carbon and hydrogen isotope fractionation resulting from anaerobic methane oxidation. *Glob. Biogeochem. Cycles* **1988**, *2*, 279–288. [CrossRef]

21. Holler, T.; Wegener, G.; Knittel, K.; Boetius, A.; Brunner, B.; et al. Substantial ^{13}C/^{12}C and D/H fractionation during anaerobic oxidation of methane by marine consortia enriched in vitro. *Environ. Microbiol. Rep.* **2009**, *1*, 370–376. [CrossRef]

22. Games, L.M.; Hayes, J.M.; Gunsalus, R.P. Methane-producing bacteria: Natural fractionations of the stable carbon isotopes. *Geochim. Cosmochim. Acta* **1978**, *42*, 1295–1297. [CrossRef]

23. Fuchs, G.; Thauer, R.; Ziegler, H.; Stichler, W. Carbon isotope fractionation by *Methanobacterium thermoautotrophicum*. *Arch. Microbiol.* **1979**, *120*, 135–139. [CrossRef]

24. Belyaev, S.S.; Wolkin, R.; Kenealy, W.R.; DeNiro, M.J.; Epstein, S.; Zeikus, J.G. Methanogenic bacteria from the Bondyuzhskoe oil field: General characterization and analysis of stable-carbon isotopic fractionation. *Appl. Environ. Microbiol.* **1983**, *45*, 691–697.

25. Balabane, M.; Galimov, E.; Hermann, M.; Létolle, R. Hydrogen and carbon isotope fractionation during experimental production of bacterial methane. *Org. Geochem.* **1987**, *11*, 115–119. [CrossRef]

26. Krzycki, J.A.; Kenealy, W.R.; DeNiro, M.J.; Zeikus, J.G. Stable carbon isotope fractionation by *Methanosarcina barkeri* during methanogenesis from acetate, methanol or carbon dioxide-hydrogen. *Appl. Environ. Microbiol.* **1987**, *53*, 2597–2599. [PubMed]

27. Gelwicks, J.T.; Risatti, J.B.; Hayes, J.M. Carbon isotope effects associated with aceticlastic methanogenesis. *Appl. Environ. Microbiol.* **1994**, *60*, 467–472. [PubMed]

28. Botz, R.; Pokojski, H.D.; Schmitt, M.; Thomm, M. Carbon isotope fractionation during bacterial methanogenesis by CO_2 reduction. *Org. Geochem.* **1996**, *25*, 255–262. [CrossRef]

29. Summons, R.E.; Franzmann, P.D.; Nichols, P.D. Carbon isotopic fractionation associated with methylotrophic methanogenesis. *Org. Geochem.* **1998**, *28*, 465–475. [CrossRef]

30. Valentine, D.L.; Chidthaisong, A.; Rice, A.; Reeburgh, W.S.; Tyler, S.C. Carbon and hydrogen isotope fractionation by moderately thermophilic methanogens. *Geochim. Cosmochim. Acta* **2004**, *68*, 1571–1590. [CrossRef]

31. Londry, K.L.; Dawson, K.G.; Grover, H.D.; Summons, R.E.; Bradley, A.S. Stable carbon isotope fractionation between substrates and products of *Methanosarcina barkeri*. *Org. Geochem.* **2008**, *39*, 608–621. [CrossRef]

32. Okumura, T.; Kawagucci, S.; Saito, Y.; Matsui, Y.; Takai, K.; Imachi, H. Hydrogen and carbon isotope systematics in hydrogenotrophic methanogenesis under H_2-limited and H_2-enriched conditions: Implications for the origin of methane and its isotopic diagnosis. *Prog. Earth Planet. Sci.* **2016**, *3*, 1–19. [CrossRef]

33. Miller, H.M.; Chaudhry, N.; Conrad, M.E.; Bill, M.; Kopf, S.H.; Templeton, A.S. Large carbon isotope variability during methanogenesis under alkaline conditions. *Geochim. Cosmochim. Acta* **2018**, *237*, 18–31. [CrossRef]

34. Horita, J. Carbon isotope exchange in the system CO_2-CH_4 at elevated temperatures. *Geochim. Cosmochim. Acta* **2001**, *65*, 1907–1919. [CrossRef]

35. Sugimoto, A.; Wada, E. Carbon isotopic composition of bacterial methane in a soil incubation experiment: Contributions of acetate and CO_2/H_2. *Geochim. Cosmochim. Acta* **1993**, *57*, 4015–4027. [CrossRef]

36. Whiticar, M.J.; Faber, E.; Schoell, M. Biogenic methane formation in marine and freshwater environments: CO_2 reduction vs. acetate fermentation—Isotope evidence. *Geochim. Cosmochim. Acta* **1986**, *50*, 693–709. [CrossRef]

37. Jørgensen, B.B.; D'Hondt, S. A starving majority deep beneath the seafloor. *Science* **2006**, *314*, 932–934. [CrossRef] [PubMed]

38. Holler, T.; Wegener, G.; Niemann, H.; Deusner, C.; Ferdelman, T.G.; Boetius, A.; Brunner, B.; Widdel, F. Carbon and sulfur back flux during anaerobic microbial oxidation of methane and coupled sulfate reduction. *Proc. Natl. Acad. Sci. USA* **2011**, *108*, 1484–1490. [CrossRef]

39. Yoshinaga, M.Y.; Holler, T.; Goldhammer, T.; Wegener, G.; Pohlman, J.W.; Brunner, B.; Kuypers, M.M.M.; Hinrichs, K.U.; Elvert, M. Carbon isotope equilibration during sulphate-limited anaerobic oxidation of methane. *Nat. Geosci.* **2014**, *7*, 190–194. [CrossRef]

40. Urey, H.C.; Greiff, L.J. Isotopic exchange equilbria. *J. Am. Chem. Soc.* **1935**, *57*, 321–327. [CrossRef]

41. Giggenbach, W.F. Carbon-13 exchange between CO_2 and CH_4 under geothermal conditions. *Geochim. Cosmochim. Acta* **1982**, *46*, 159–165. [CrossRef]

42. Richet, P.; Bottinga, Y.; Javay, M. A review of hydrogen, carbon, nitrogen, oxygen, sulphur, and chlorine stable isotope fractionation among gaseous molecules. *Ann. Rev. Earth Planet. Sci.* **1977**, *5*, 65–110. [CrossRef]

43. Bottinga, Y. Calculated fractionation factors for carbon and hydrogen isotope exchange in the system calcite-carbon dioxide-graphite-methane-hydrogen-water vapour. *Geochim. Cosmochim. Acta* **1969**, *33*, 49–64. [CrossRef]

44. Penning, H.; Plugge, C.M.; Galand, P.E.; Conrad, R. Variation of carbon isotope fractionation in hydrogenotrophic methanogenic microbial cultures and environmental samples at different energy status. *Glob. Chang. Biol.* **2005**, *11*, 2103–2113. [CrossRef]

45. Moran, J.J.; House, C.H.; Freeman, K.H.; Ferry, J.G. Trace methane oxidation studied in several Euryarchaeota under diverse conditions. *Archaea* **2005**, *1*, 303–309. [CrossRef] [PubMed]

46. Takai, K.; Nakamura, K.; Toki, T.; Tsunogai, U.; Miyazaki, M.; Miyazaki, J.; Hirayama, H.; Nakagawa, S.; Nunoura, T.; Horikoshi, K. Cell proliferation at 122 °C and isotopically heavy CH_4 production by a hyperthermophilic methanogen under high-pressure cultivation. *Proc. Natl. Acad. Sci. USA* **2008**, *105*, 10949–10954. [CrossRef] [PubMed]

47. Adam, P.S.; Borrel, G.; Gribaldo, S. Evolutionary history of carbon monoxide dehydrogenase/acetyl-CoA synthase, one of the oldest enzymatic complexes. *Proc. Natl. Acad. Sci. USA* **2018**, *115*, 1166–1173. [CrossRef] [PubMed]

48. Martin, W.; Russel, M.J. On the origin of biochemistry at an alkaline hydrothermal vent. *Phil. Trans. R. Soc. B* **2007**, *362*, 1887–1925. [CrossRef] [PubMed]

49. Weiss, M.C.; Sousa, F.L.; Mrnjavac, N.; Neukirchen, S.; Roettger, M.; Nelson-Sathi, S.; Martin, W.F. The physiology and habitat of the last universal common ancestor. *Nat. Microbiol.* **2016**, *1*, 16116. [CrossRef]

50. Berghuis, B.A.; Yu, F.B.; Schulz, F.; Blainey, P.C.; Woyke, T.; Quake, S.R. Hydrogenotrophic methanogenesis in archaeal phylum Verstraetearchaeota reveals the shared ancestry of all methanogens. *Proc. Natl. Acad. Sci. USA* **2019**, *116*, 5037–5044. [CrossRef]

51. Kellermann, M.Y.; Wegener, G.; Elvert, M.; Yoshinaga, M.Y.; Lin, Y.S.; Thomas, H.; Prieto Mollar, X.; Knittel, K.; Hinrichs, K.U. Autotrophy as a predominant mode of carbon fixation in anaerobic methane-oxidizing microbial communities. *Proc. Natl. Acad. Sci. USA* **2012**, *109*, 19321–19326. [CrossRef]

52. Alperin, M.J.; Hoehler, T.M. Anaerobic methane oxidation by Archaea/sulfate-reducing Bacteria aggregates: 2. isotopic constraints. *Am. J. Sci.* **2009**, *309*, 958–984. [CrossRef]

53. Skennerton, C.T.; Chourey, K.; Iyer, R.; Hettich, R.L.; Tyson, G.W.; Orphan, V.J. Methane-fueled syntrophy through extracellular electron transfer: Uncovering the genomic traits conserved within diverse bacterial partners of anaerobic methanotrophic Archaea. *mBio* **2017**, *8*, e00530-17. [CrossRef]

54. Thauer, R.K.; Kaster, A.K.; Seedorf, H.; Buckel, W.; Hedderich, R. Methanogenic archaea: Ecologically relevant differences in energy conservation. *Nat. Rev.* **2008**, *6*, 579–591. [CrossRef]

55. Meister, P.; Liu, B.; Khalili, A.; Böttcher, M.E.; Jørgensen, B.B. Factors controlling the carbon isotope composition of dissolved inorganic carbon and methane in marine porewater: An evaluation by reactive-transport modelling. *J. Mar. Syst.* **2019**, *200*, 103227. [CrossRef]

56. Blair, N.E.; Carter, W.D., Jr. The carbon isotope biogeochemistry of acetate from a methanogenic marine sediment. *Geochim. Cosmochim. Acta* **1992**, *56*, 1247–1258. [CrossRef]

57. Hinrichs, K.U.; Summons, R.E.; Orphan, V.; Sylva, S.P.; Hayes, J.M. Molecular and isotopic analyses of anaerobic methane-oxidizing communities in marine sediments. *Org. Geochem.* **2000**, *31*, 1685–1701. [CrossRef]

58. Pancost, R.D.; Sinninghe Damsté, J.S.; de Lint, S.; van der Maarel, M.J.; Gottschal, J.C. The Medinaut Shipboard Scientific Party Biomarker evidence for widespread anaerobic methane oxidation in Mediterranean sediments by a consortium of methanogenic Archaea and bacteria. *Appl. Environ. Microbiol.* **2000**, *66*, 1126–1132. [CrossRef]

59. Contreras, S.; Meister, P.; Liu, B.; Prieto-Mollar, X.; Hinrichs, K.U.; Khalili, A.; Ferdelman, T.G.; Kuypers, M.; Jørgensen, B.B. Strong glacial-interglacial variation of sub-seafloor microbial activity on the Peruvian shelf. *Proc. Natl. Acad. Sci. USA* **2013**, *110*, 18098–18103. [CrossRef]

60. Carr, S.A.; Schubotz, F.; Dunbar, R.B.; Mills, C.T.; Dias, R.; Summons, R.E.; Mandernack, K.W. Acetoclastic Methanosaeta are dominant methanogens in organic-rich Antarctic marine sediments. *Int. Soc. Microb. Ecol. J.* **2018**, *12*, 330–342. [CrossRef]

61. McCollom, T.M.; Bach, W. Thermodynamic constraints on hydrogen generation during serpentinization of ultramafic rocks. *Geochim. Cosmochim. Acta* **2009**, *73*, 856–875. [CrossRef]

62. Meister, P.; Wiedling, J.; Lott, C.; Bach, W.; Kuhfuß, H.; Wegener, G.; Böttcher, M.E.; Deusner, C.; Lichtschlag, A.; Bernasconi, S.M.; et al. Anaerobic methane oxidation inducing carbonate precipitation at abiogenic methane seeps in the Tuscan archipelago (Italy). *PLoS ONE* **2018**, *13*, e0207305. [CrossRef]

63. Ijiri, A.; Harada, N.; Hirota, A.; Tsunogai, U.; Ogawa, N.O.; Itaki, T.; Khim, B.K.; Uchida, M. Biogeochemical processes involving acetate in sub-seafloor sediments from the Bering Sea shelf break. *Org. Geochem.* **2012**, *48*, 47–55. [CrossRef]

64. Zeebe, R.E. Modeling CO_2 chemistry, $\delta^{13}C$, and oxidation of organic carbon and methane in sediment porewater: Implications for paleo-proxies in benthic foraminifera. *Geochim. Cosmochim. Acta* **2007**, *71*, 3238–3256. [CrossRef]

65. Drake, H.; Åström, M.E.; Heim, C.; Broman, C.; Åström, J.; Whitehouse, M.; Ivarsson, M.; Siljeström, S.; Sjövall, P. Extreme 13 C depletion of carbonates formed during oxidation of biogenic methane in fractured granite. *Nat. Commun.* **2015**, *6*, 7020. [CrossRef] [PubMed]

66. Drake, H.; Heim, C.; Roberts, N.M.W.; Zack, T.; Tillberg, M.; Broman, C.; Ivarsson, M.; Whitehouse, M.J.; Åström, M.E. Isotopic evidence for microbial production and consumption of methane in the upper continental crust throughout the Phanerozoic eon. *Earth Planet. Sci. Lett.* **2017**, *470*, 108–118. [CrossRef]

67. Nissenbaum, A.; Presley, B.J.; Kaplan, I.R. Early diagenesis in a reducing fjord, Saanich Inlet, British Columbia–I. Chemical and isotopic changes in major components of interstitial water. *Geochim. Cosmochim. Acta* **1972**, *36*, 1007–1027. [CrossRef]

68. Paull, C.K.; Lorenson, T.D.; Borowski, W.S.; Ussler, W., III; Olsen, K.; Rodriguez, N.M. Isotopic composition of CH_4, CO_2 species, and sedimentary organic matter within samples from the Blake Ridge: Gas source implications. *Proc. ODP Sci. Results* **2000**, *164*, 67–78.

69. Chatterjee, S.; Dickens, G.R.; Bhatnagar, G.; Chapman, W.G.; Dugan, B.; Snyder, G.T.; Hirasaki, G.J. Pore water sulfate, alkalinity, and carbon isotope profiles in shallow sediment above marine gas hydrate systems: A numerical modeling perspective. *J. Geophys. Res.* **2011**, *116*, B9. [CrossRef]

70. Pohlman, J.W.; Ruppel, C.; Hutchinson, D.R.; Coffin, R.B. Assessing sulfate reduction and methane cycling in a high salinity pore water system in the northern Gulf of Mexico. *Mar. Pet. Geol.* **2008**, *25*, 942–951. [CrossRef]

71. Wu, Z.; Liu, B.; Escher, P.; Kowalski, N.; Böttcher, M.E. Carbon diagenesis in different sedimentary environments of the subtropical Beibu Gulf, South China Sea. *J. Mar. Syst.* **2018**, *186*, 68–84. [CrossRef]

72. Chuang, P.C.; Yang, T.F.; Wallmann, K.; Matsumoto, R.; Hu, C.Y.; Chen, H.W.; Lin, S.; Sun, C.H.; Li, H.C.; Wang, Y.; et al. Carbon isotope exchange during anaerobic oxidation of methane (AOM) in sediments of the northeastern South China Sea. *Geochim. Cosmochim. Acta* **2019**, *246*, 138–155. [CrossRef]

73. Meister, P.; Liu, B.; Ferdelman, T.G.; Jørgensen, B.B.; Khalili, A. Control of sulphate and methane distributions in marine sediments by organic matter reactivity. *Geochim. Cosmochim. Acta* **2013**, *104*, 183–193. [CrossRef]

74. Dale, A.W.; Aguilera, D.R.; Regnier, P.; Fossing, H.; Knab, N.J.; Jørgensen, B.B. Seasonal dynamics of the depth and rate of anaerobic oxidation of methane in Aarhus Bay (Denmark) sediments. *J. Mar. Res.* **2008**, *66*, 127–155. [CrossRef]

75. Mogollón, J.M.; L'Heureux, I.; Dale, A.W.; Regnier, P. Methane gas-phase dynamics in marine sediments: A model study. *Am. J. Sci.* **2009**, *309*, 189–220. [CrossRef]

76. Smrzka, D.; Kraemer, S.M.; Zwicker, J.; Birgel, D.; Fischer, D.; Kasten, S.; Goedert, J.L.; Peckmann, J. Constraining silica diagenesis in methane-seep deposits. *Palaeogeogr. Palaeoclim. Palaeoecol.* **2015**, *420*, 13–26. [CrossRef]

77. Boudreau, B.P. The physics of bubbles in surficial, soft, cohesive sediments. *Mar. Pet. Geol.* **2012**, *38*, 1–18. [CrossRef]

78. Dale, A.W.; Van Cappellen, P.; Aguilera, D.R.; Regnier, P. Methane efflux from marine sediments in passive and active margins: Estimations from bioenergetic reaction-transport simulations. *Earth Planet. Sci. Lett.* **2008**, *265*, 329–344. [CrossRef]

79. Zeebe, R.E. On the molecular diffusion coefficients of dissolved CO_2, HCO_3^-, and CO_3^{2-}, and their dependence on isotopic mass. *Geochim. Cosmochim. Acta* **2011**, *75*, 2483–2498. [CrossRef]

80. Borowski, W.S.; Paull, C.K.; Ussler, W., III. Global and local variations of interstitial sulfate gradients in deep-water, continental margin sediments: Sensitivity to underlying methane and gas hydrates. *Mar. Geol.* **1999**, *159*, 131–154. [CrossRef]

81. Kennett, J.P.; Cannariato, K.G.; Hendy, I.L.; Behl, R.J. Carbon isotopic evidence for methane hydrate instability during Quaternary Interstadials. *Science* **2000**, *288*, 128–133. [CrossRef]

82. Whiticar, M.J. Carbon and hydrogen isotope systematics of bacterial formation and oxidation of methane. *Chem. Geol.* **1999**, *161*, 291–314. [CrossRef]

83. Baker, P.A.; Burns, S.J. Occurrence and formation of dolomite in organic-rich continental margin sediments. *Am. Assoc. Pet. Geol. Bull.* **1985**, *69*, 1917–1930.

84. Meister, P.; Gutjahr, M.; Frank, M.; Bernasconi, S.; McKenzie, J.A.; Vasconcelos, C. Dolomite formation within the methanogenic zone induced by tectonically-driven fluids in the Peru accretionary prism. *Geology* **2011**, *39*, 563–566. [CrossRef]

85. Meister, P. Two opposing effects of sulfate reduction on calcite and dolomite precipitation in marine, hypersaline and alkaline environments. *Geology* **2013**, *41*, 499–502. [CrossRef]

86. Moore, T.S.; Murray, R.W.; Kurtz, A.C.; Schrag, D.P. Anaerobic methane oxidation and the formation of dolomite. *Earth Planet. Sci. Lett.* **2004**, *229*, 141–154. [CrossRef]

87. Ussler, W., III; Paull, C.K. Rates of anaerobic oxidation of methane and authigenic carbonate mineralization in methane-rich deep-sea sediments inferred from models and geochemical profiles. *Earth Planet. Sci. Lett.* **2008**, *266*, 271–287. [CrossRef]

88. Wallmann, K.; Aloisi, G.; Haeckel, M.; Tishchenko, P.; Pavlova, G.; Greinert, J.; Kutterolf, S.; Eisenhauer, A. Silicate weathering in anoxic marine sediments. *Geochim. Cosmochim. Acta* **2008**, *72*, 3067–3090. [CrossRef]

89. Wehrmann, L.M.; Ockert, C.; Mix, A.; Gussone, N.; Teichert, B.M.A.; Meister, P. Multiple onset of methanogenic zones, diagenetic dolomite formation, and silicate alteration under varying organic carbon deposition in Bering Sea sediments (Bowers Ridge, IODP Exp. 323 Site U1341). *Deep Sea Res. II* **2016**, *125*, 117–132. [CrossRef]

90. Turner, J.V. Kinetic fractionation of carbon-13 during calcium carbonate precipitation. *Geochim. Cosmochim. Acta* **1982**, *46*, 1183–1191. [CrossRef]

91. Deines, P.; Langmuir, D.; Harmon, R.S. Stable Carbon isotope ratios and the existence of a gas phase in the evolution of carbonate ground waters. *Geochim. Cosmochim. Acta* **1974**, *38*, 1147–1164. [CrossRef]

92. Ohmoto, H.; Rye, R.O. Isotopes of sulphur and carbon. In *Geochemistry of Hydrothermal Deposits*; Barnes, H.L., Ed.; John Wiley & Sons: Hoboken, NJ, USA, 1979; pp. 509–567.

93. Golyshev, S.I.; Padalko, N.L.; Pechenkin, S.A. Fractionation of stable oxygen and carbon isotopes in carbonate systems. *Geochem. Int.* **1981**, *18*, 85–99.

94. Mook, W.G.; Bommerson, J.C.; Staverman, W.H. Carbon isotope fractionation between dissolved bicarbonate and gaseous carbon dioxide. *Earth Planet. Sci. Lett.* **1974**, *22*, 169–176. [CrossRef]

95. Michaelis, J.; Usdowski, E.; Menschel, G. Partitioning of ^{13}C and ^{12}C on the degassing of CO_2 and the precipitation of calcite—Rayleigh-type fractionation and a kinetic model. *Am. J. Sci.* **1985**, *285*, 318–327. [CrossRef]

96. Meister, P.; Bernasconi, S.M.; Aiello, I.W.; Vasconcelos, C.; McKenzie, J.A. Depth and controls of Ca-rhodochrosite precipitation in bioturbated sediments of the Eastern Equatorial Pacific, ODP Leg 201, Site 1226 and DSDP Leg 68, Site 503. *Sedimentology* **2009**, *56*, 1552–1568. [CrossRef]

97. Kasina, M.; Bock, S.; Wuerdemann, H.; Pudlo, D.; Picard, A.; Lichtschlag, A.; März, C.E.; Wagenknecht, L.; Wehrmann, L.M.; Vogt, C.; et al. Mineralogical and geochemical analysis of Fe-phases in drill-cores from the Triassic Stuttgart Formation at Ketzin CO_2 storage site before CO_2 arrival. *Environ. Earth Sci.* **2017**, *161*, 161. [CrossRef]

98. Bernasconi, S.M. Geochemical and microbial controls on dolomite formation in anoxic environments: A case study from the Middle Triassic (Ticino, Switzerland). In *Contributions to Sedimentology*; E. Schweizerbart Science Publishers: Stuttgart, Germany, 1994; Volume 19, pp. 1–109.

99. Meister, P.; McKenzie, J.A.; Bernasconi, S.M.; Brack, P. Dolomite formation in the shallow seas of the Alpine Triassic. *Sedimentology* **2013**, *60*, 270–291. [CrossRef]

100. Heimhofer, U.; Meister, P.; Bernasconi, S.M.; Ariztegui, D.; Martill, D.M.; Rios-Netto, A.M.; Schwark, L. Isotope and elemental geochemistry of black shale-hosted fossiliferous concretions (Early Cretaceous Santana Formation, Brazil). *Sedimentology* **2017**, *64*, 150–167. [CrossRef]

101. Thang, N.M.; Brüchert, V.; Formolo, M.; Wegener, G.; Ginters, L.; Jørgensen, B.B.; Ferdelman, T.G. The impact of sediment and carbon fluxes on the biogeochemistry of methane and sulfur in littoral Baltic Sea sediments (Himmerfjärden, Sweden). *Estuar. Coasts* **2013**, *36*, 98–115. [CrossRef]

102. Jørgensen, N.O. Methane-derived carbonate cementation of marine sediments from the Kattegat, Denmark: Geochemical and geological evidence. *Mar. Geol.* **1992**, *103*, 1–13. [CrossRef]

103. Pisciotto, K.A.; Mahoney, J.J. Isotopic survey of diagenetic carbonates, Deep Sea Drilling Project Leg 63. In *Initial Reports of the Deep Sea Drilling Project*; Yeats, R.S., Haq, B.U., Eds.; U.S. Govt. Printing Office: Washington, DC, USA, 1981; Volume 63, pp. 595–609.

104. Kelts, K.; McKenzie, J.A. Diagenetic dolomite formation in Quaternary anoxic diatomaceous muds of Deep Sea Drilling Project Leg 64, Gulf of California. In *Init. Repts. DSDP*; Curray, J.R., Moore, D.G., Eds.; U.S. Govt. Printing Office: Washington, DC, USA, 1982; Volume 64, pp. 553–569.

105. Murata, K.J.; Friedman, I.; Madsen, B.H. *Isotopic Composition of Diagenetic Carbonates in Marine Miocene Formations of California and Oregon*; USGS Prof. Paper; U.S. Govt. Printing Office: Washington, DC, USA, 1969; p. 24.

106. Rodriguez, N.M.; Paull, C.K.; Borowski, W.S. Zonation of authigenic carbonates within gas hydrate-bearing sedimentary sections on the blake ridge: Offshore southeastern North America. In *Proc. ODP, Sci. Results*; Paull, C.K., Matsumoto, R., Wallace, P.J., Dillon, W.P., Eds.; Ocean Drilling Program: College Station, TX, USA, 2000; Volume 164, pp. 301–312.

107. Compton, J.S. Sediment composition and precipitation of dolomite and pyrite in the Neogene Monterey and Sisquoc Formations, Santa Maria Basin area, California. In *Sedimentology and Geochemistry of Dolostones*; Shukla, V., Baker, P.A., Eds.; SEPM Spec. Pub.: Tulsa, OK, USA, 1988; Volume 43, pp. 53–64.

108. Meister, P.; Bernasconi, S.; McKenzie, J.A.; Vasconcelos, C. Sea-level changes control diagenetic dolomite formation in hemipelagic sediments of the Peru Margin. *Mar. Geol.* **2008**, *252*, 166–173. [CrossRef]

109. Alt, J.C.; Teagle, D.A.H. The uptake of carbon during alteration of ocean crust. *Geochim. Cosmochim. Acta* **1999**, *63*, 1527–1535. [CrossRef]

110. Dickens, G.R. A blast of gas in the latest Paleocene: Simulating first-order effects of massive dissociation of oceanic methane hydrate. *Geology* **1997**, *25*, 259–262. [CrossRef]

111. Knoll, A.H.; Hayes, J.M.; Kaufmann, A.J.; Swett, K.; Lambert, I.B. Secular variation in carbon isotope ratios from Upper Proterozoic successions of Svalbard and East Greenland. *Nature* **1986**, *321*, 832–838. [CrossRef]

112. Payne, J.L.; Lehrmann, D.J.; Wei, J.; Orchard, M.J.; Schrag, D.P.; Knoll, A.H. Large perturbations of the carbon cycle during recovery from the end-Permian extinction. *Science* **2004**, *305*, 506–509. [CrossRef] [PubMed]

113. Wefer, G.; Heinze, P.M.; Berger, W.H. Clues to ancient methane release. *Nature* **1994**, *369*, 282. [CrossRef]

114. Louis-Schmid, B.; Rais, P.; Logvinovich, D.; Bernasconi, S.M.; Weissert, H. Impact of methane seeps on the local carbon-isotope record: A case study from a Late Jurassic hemipelagic section. *Terra Nova* **2007**, *19*, 259–265. [CrossRef]

115. Schidlowski, M.; Matzigkeit, U.; Krumbein, W.E. Superheavy organic carbon from hypersaline microbial mats: Assimilatory pathway and geochemical implications. *Naturwissenschaften* **1984**, *71*, 303–308. [CrossRef]
116. Hayes, J.M.; Waldbauer, J., Jr. The carbon cycle and associated redox processes through time. *Philos. Trans. R. Soc. B Biol. Sci.* **2006**, *361*, 931–950. [CrossRef]
117. Birgel, D.; Meister, P.; Lundberg, R.; Horat, T.; Bontognali, T.; Bahniuk, A.; Rezende, C.; Vasconcelos, C.; McKenzie, J.A. Methanogenesis produces strong ^{13}C-enrichment in stromatolites of Lagoa Salgada, Brazil: A modern analogue for Palaeo-/Neoproterozoic stromatolites? *Geobiology* **2015**, *13*, 245–266. [CrossRef]

Review

Instant Attraction: Clay Authigenesis in Fossil Fungal Biofilms

Therese Sallstedt [1,*], Magnus Ivarsson [1,2], Henrik Drake [3] and Henrik Skogby [4]

[1]   Department of Paleobiology, Swedish Museum of Natural History, 11418 Stockholm, Sweden
[2]   Department of Biology, University of Southern Denmark, Campusvej 55, Odense M, 5230 Odense, Denmark
[3]   Department of Biology and Environmental Science, Linnaeus University, 39182 Kalmar, Sweden
[4]   Department of Geology, Swedish Museum of Natural History, 11418 Stockholm, Sweden
*   Correspondence: therese.sallstedt@nrm.se

Received: 25 June 2019; Accepted: 21 August 2019; Published: 24 August 2019

Abstract: Clay authigenesis associated with the activity of microorganisms is an important process for biofilm preservation and may provide clues to the formation of biominerals on the ancient Earth. Fossilization of fungal biofilms attached to vesicles or cracks in igneous rock, is characterized by fungal-induced clay mineralization and can be tracked in deep rock and deep time, from late Paleoproterozoic (2.4 Ga), to the present. Here we briefly review the current data on clay mineralization by fossil fungal biofilms from oceanic and continental subsurface igneous rock. The aim of this study was to compare the nature of subsurface fungal clays from different igneous settings to evaluate the importance of host rock and ambient redox conditions for clay speciation related to fossil microorganisms. Our study suggests that the most common type of authigenic clay associated with pristine fossil fungal biofilms in both oxic (basaltic) and anoxic (granitic) settings are montmorillonite-like smectites and confirms a significant role of fungal biofilms in the cycling of elements between host rock, ocean and secondary precipitates. The presence of life in the deep subsurface may thus prove more significant than host rock geochemistry in directing the precipitation of authigenic clays in the igneous crust, the extent of which remains to be fully understood.

Keywords: clay authigenesis; fossil fungi; igneous crust; cryptoendoliths; subseafloor habitats

---

1. Introduction

The formation of authigenic minerals is a fundamental aspect of the interaction between microbial life and the ambient environment [1–10]. Throughout major parts of Earth history, clay minerals and low-crystalline amorphous aluminosilicate-phases dominated by a Fe-Al-Si composition, have been associated with biology, often in the form of clay mineralized prokaryotic [4,5,7,11,12] or eukaryotic cells and biofilms [13–19]. The close association and interconnection between microbes and authigenic clays have been demonstrated in vitro from laboratory culture experiments [20,21], but also from natural sedimentary environments, lacustrine sediments [2,4,5,22], hydrothermal areas [23], terrestrial rock habitats [24] and within the subsurface igneous biosphere of both continents and oceans [15–18,25]. Over the past decade, several studies have emphasized the importance of a subsurface crustal biota and its prominent role in element cycling, rock weathering and precipitation of secondary minerals [15–19,25–28]. Because of the vastness and relative inaccessibility of the deep biosphere in the oceanic and continental crust, however, more work is needed in order to fully comprehend and characterize the specific nature of life and trophic systems in the subsurface crust.

The current knowledge of deep biosphere-ecology is primarily based on traces of presumed microbial activity within the glassy rim of basaltic rocks [29–31], as well as body fossils within subseafloor igneous settings [13–19,25,32–39]. In particular, fungal filaments and biofilms preserved by authigenic clays seems to represent a large fraction of the deep biosphere fossil biotas [16–19,33–36,38].

While the presence of fossil fungi from the oceanic crust have been well established during the last decade, new research confirms a significant fungal presence also within the anoxic continental crust [25,39,40]. In drill cores from the Laxemar site, near Äspö hard rock laboratory in Sweden, partly carbonaceous and clay mineralized fungi have been identified within a granite hosted vein-system, associated with reducing minerals such as pyrite, which indicates that the granite-dwelling fungal communities had a largely anaerobic lifestyle [25].

The mode of clay encrustation within microorganisms in deep subsurface habitats resembles that identified within bacteria and prokaryotic biofilms in lake sediments, previously described by Ferris et al. [1,3], Konhauser and Urrutia [7] and Konhauser et al. [12], among others. This similarity not only suggests a common genesis, but, more importantly, highlights the affinity of microbial cells and their extracellular polymeric substances (EPS) to attract clay-forming elements leading to the formation of authigenic minerals. An important characteristic, therefore, of all cellular surfaces and their EPS, is the high reactivity and potential to adsorb, retain and exchange chemical species, including elements essential for microbial metabolism [41,42]. These characteristics can lower activation-energy and thus invoke nucleation of secondary minerals in association with microbial walls or sheaths [8,43–47]. This is also true of phyllosilicate minerals per se, and both clays and the reactive cell or biofilm surfaces, can therefore be considered important targets for understanding biogeochemical element cycling and the transfer of elements between rock, hydrosphere and the biota [48].

Most previous studies concerning microbial clay mineralization have focused on the role of prokaryotic microorganisms in shallow settings, such as soils, riverine, ocean and geothermal-sediments, as well as terrestrial flood basalts [1,2,4,5,11,23]. However, the increased understanding and characterization of extreme oligotrophic habitats in igneous rock calls for a closer examination of clay authigenesis within the subsurface crust. Understanding fossilization processes and the timing and mode of microbial preservation, can give important clues to the nature of life and its interaction with the environment. Therefore, the overall aim of this study is to add insight into the unexplored subject of fungal clay fossilization, by comparing the geochemistry of fungal clay precipitates from various igneous subsurface localities. New X-ray powder diffraction (XRD) and Mössbauer Spectroscopy data from Pacific Ocean pillow basalt and two anoxic fractured granite cores from Laxemar, Sweden, was added in order to determine and compare clay chemistry and Fe oxidation states between anoxic and oxic settings.

Our review suggests that intrinsic factors such as the structure of fungal biofilms may have a larger effect on the mineralogy of fungal clays compared to abiotically precipitated clays forming solely in equilibrium with surrounding source rock and fluids.

2. Methods

2.1. X-Ray Powder Diffraction

We used X-ray powder diffraction in order to analyse the mineralogical composition of authigenic clays from subsurface anoxic granite rocks and oxygenated basaltic seafloor rocks. The samples were removed from different depths and carefully extracted from split drill-cores using sterile forceps. The first sample consisted entirely of fungal filaments from a well-preserved biofilm (granite sample XLX09). A second, darker, clay phase represented a less well-preserved biofilm matrix (granite sample XCX04). A third clay sample (basalt sample 1205A-024R) associated with fungal filaments was scraped from a split drill core originating from the Pacific Ocean Nintoku seamount. The XRD patterns were collected using a Panalytical X'pert powder diffractometer (PANalytical B.V., Almelo, The Netherlands) equipped with an X'celerator silicon-strip detector. The range 5–80° (2θ) was scanned with a step-size of 0.017° using a sample spinner with finely ground sample material mounted on a background-free silicon holder. All samples were mixed with dH$_2$O and run a second time, to evaluate the potential effect of clay swelling on the XRD spectra.

2.2. Mössbauer Spectroscopy

To characterize the redox state of Fe within the three clay samples investigated with XRD described above, Mössbauer spectra were measured on samples XLX09 (anoxic granite), XCX04 (anoxic granite) and 1205A-024R (oxic Pacific Ocean basalt) using a conventional spectrometer system operated in constant-acceleration mode. Due to limited amounts of sample material, a ^{57}Co rhodium matrix point-source with a nominal activity of 10 mCi was used. The absorbers were prepared by a mixture of sample material (<1 mg) and thermoplastic resin, which was shaped to ca. 1-mm sized cylinders under mild heating (<100 °C), placed on strip tape and positioned close to the source. The spectra were collected at room-temperature over the velocity range −4.2 to +4.2 mm^{-1} distributed over 1024 channels and were calibrated against an a-Fe foil before folding and spectral fitting using the software MossA [49].

3. Microbial Clay Mineralization

Several studies emphasize the importance of microbial cells in the precipitation of iron rich minerals, in particular authigenic clays [1–5,7,22,50,51]. Studies focused on prokaryotic mineralization show that biogenic clay authigenesis occurs through a two-step process, where the initial encrustation of iron surrounding cells or filaments is followed by a secondary step of Al-Si complexion [7]. This process can result in the formation of amorphous aluminosilicate phases surrounding the microbial cell wall or EPS. Microorganisms can thus become completely or partly mineralized, with the newly formed clay grains tangentially attached to the cell wall [7]. An early cellularly bound iron-phase seems to act as a ligand that initiates subsequent precipitation of poorly crystallized aluminosilicates, potentially in the form of a precursor gel, which eventually mature into more crystalline clay phases, with the potential for fossil record preservation [13–19,25,32,52,53].

While clay formation may occur as a result of abiotic processes, for example chemical weathering resulting from interaction between oxygenated seawater and reducing basaltic rock, or hydrothermal water circulation [54], the presence of microbial cells and the biofilm EPS, appear to play a significant role in clay authigenesis [1–3,7,51–53]. One way to differentiate between biotically precipitated clay phases and those formed by abiotic processes, is to compare the degree of heterogeneity and crystallinity within the clay phases [7]. For example, Si has been proposed to play a larger role in the structure of abiotic clays, compared to biogenic bacterial precipitates, which are characterized by a lower Si concentration, smaller grain size, and frequently a more amorphous structure [7]. Further, the ratio of various elements within the clays has a large impact on the physical stability and geochemical characteristics of the mineral [55]. In smectites, for example, the amount of structural Fe affects the swelling ability of the clay phase; and Fe-rich smectites have a reduced expansion ability compared with lower Fe smectites [56]. The association and interaction between microbes and clay-forming elements described above appears to occur in many types of environments. For example, Sanchez-Navaz et al. [52] described the mineralization of iron-rich, poorly ordered and amorphous smectites in association with apatite grains within phosphatic stromatolites from the Upper Jurassic. They suggest that the presence of authigenic smectites in microbially produced stromatolite lamina are the result of a biogenically produced precursor-phase rich in Fe-Si-Al, which matures into a smectite phase during early diagenesis [52]. Here, the presence of the microbial biofilm EPS presumably acts as a catalyst for the attraction of cations and other chemical species forming the precursor gel-like phase, which subsequently develops into a clay like aluminosilicates [52]. Further, in oligotrophic granite at the Äspö hard rock laboratory near Oskarshamn in Sweden, experiments were initiated where microorganisms from Äspö were isolated and cultured within water from the area [57]. The results showed significant precipitation of minute-sized clay particles within the presence of microbial communities, and in contrast, controlled experiments without microorganisms showed no precipitation of aluminosilicate phases [57]. Results from this experiment, therefore, suggested that either the metabolic activity, or the active surface of biological cells and matrices, were important components for the precipitation of authigenic clay phases.

Wierzchos et al. [24] similarly reported bacterial sheaths and potential fungal hyphae surrounded by Fe-rich aluminosilicate phases, in cracks or fissures within Antarctic rocks. They suggested that the combination of filaments covered in an Fe-oxyhydroxide layer and associated lamina of clay-like material, could be used as a biosignature in other similar type of settings [24]. According to the authors, it was also possible to distinguish between clays associated with well-preserved filamentous microfossils, and those of more taphonomically altered specimens; the clays surrounding dead and lysed cells tended to have a larger Fe-concentration, and in turn less Si and Al present within the clay matrix. The opposite relationship could be established among exceptionally preserved specimens [24].

Further, Fe-rich minerals close in composition to Nontronite, the Fe^{3+}-rich endmember of the smectite group, formed around microbial cells and their EPS within pillow lavas from the North East Pacific Ocean [50]. The smectite-covered communities were found in the vicinity of hydrothermal vents, which could have provided a source for Fe and Mn, which were observed in the form of oxides in association to biogenic clays. It could be established by Transmission Electron Microscopy (TEM) that the smectite phases were clearly growing from the cell walls of bacteria [50], again confirming the importance of the reactive cell surface to the nucleation and subsequent precipitation of clay-like minerals.

Similarly, within submerged laboratory cultures, Fomina and Gadd [20] examined the effect of clay on the formation of fungal pellets. Their results showed a significant thickening and lengthening of the fungal hyphae as a result of micro-sized clay particles that attached to the filaments from their suspension in the surrounding water. The absorption of clay particles onto hyphae affected the geochemical properties and permeability of the fungal pellets as well as added physical stability and strength [20].

As mentioned in the above section, the presence of an active cell surface or organic matrix like EPS, seem to play a ubiquitous part in the initial nucleation and precipitation of amorphous clay phases: The intimate relationship between microbial EPS and authigenic clays among both prokaryotes and eukaryotes is presumably due to the natural effect of microbial biofilms to attract and bind cations from the ambient fluids onto negatively charged functional groups [46,47,58–60]. Microbial EPS have often been highlighted as important agents in mineral precipitation. This has been especially noted in relation to the precipitation of calcite [44,45,47,59,61], which can precipitate within the EPS either as a result of the metabolic activity of cells, or simply by acting as a cation-trap and nucleation spot associated with acidic functional groups [45]. Important to note, however, is the dual role of EPS in mineral precipitation; because of the adsorptive properties of the organic substances, an initial inhibition of nucleation is to be expected, which is only surpassed once the cation binding-capacity of the biofilm EPS has been reached [45–47]. It is, therefore, easy to imagine a similar scenario with respect to for example Fe (e.g., [1–3,7]), in settings less supersaturated with calcite, that might instead lead to the formation of an amorphous clay-like phase in association with the EPS, depending on the ambient source of metals. Hence, extrinsic factors, such as element-availability, alkalinity, pH and saturation state of ambient fluids are all parameters that may have an effect on the type of mineralization that occurs [47].

Although a majority of studies have focused on the mineral promoting effect of prokaryotic biofilms [1–4,7], many fungal communities likewise produce large amounts of EPS, with similar cation-attracting characteristics [62–65]. Most fungi that produce EPS are aerobic, or possibly facultative anaerobs [62], and fungal production of EPS can be stimulated by increased pO_2 [65]. Ueshima and Tazaki [63], for example, showed a connection between fungal EPS and the formation of Fe-rich clay phases in the form of nontronite, which formed within the EPS, again highlighting the connection between microbes and EPS, but this time among heterotrophic eukaryotes.

One of the most important extrinsic factors that may have an effect on the production of fungal biofilm EPS, and thus in turn for clay precipitation, is the ambient pH of the host solution and growth medium [64,66]. For example, variations in pH can influence the molecular weight, as well as yield, of fungal EPS [66]. It appears as though an acidic pH generally promotes EPS production among fungi [66], and, in turn, gives a lower EPS yield, but with a higher molecular weight, compared to high pH settings that promote a low molecular weight but high EPS yield [66]. This means that the ambient environment is expected to have a large influence on fungal EPS production in general, but also on the type of mineralization that will occur. Thus, for fungal clay authigenesis, pH and EPS yield may be locally important and variable factors.

3.1. Benefits of Clay Authigenesis for Microbial Communities

Although a close spatial relationship between microbial cells, filaments and more or less well-crystallized clay phases, has been established repeatedly and from a variety of settings, the exact reason for, or nature of, the relationship between microbes and clay is still not entirely clear. While it may simply be the result of geochemical processes associated with the specific structure of active microbial surfaces, several studies suggest it may in fact be beneficial for microbial communities to surround the cells and EPS with clay phases as a way to modify the ambient surroundings favorably [67]. Because clays, like microbial expolymers, are characterized by the capacity to attract and exchange a multitude of nutrients and chemical species, this may indicate that the clay phases serve as a proximal nutrient-source for the communities [7]. This may be particularly important in extreme settings such as the highly oligotrophic igneous crust. Swelling smectite minerals such as montmorillonite, could help to stabilize the microenvironment with respect to pH, which may act beneficial on parameters such as microbial growth and respiration [7,68]. The clay can also form a protective matrix to guard viable cells against harmful effects of the surroundings, such as toxic species or damaging UV radiation [7,11,69–73]. Even many metals such as Na, K, Cu, Zn, Co, Ca, Mg, Mn, Fe, which can be considered bio-essential for fungal growth in lower concentrations, may become inhibiting in larger concentrations and can be bioremediated by means of fungal activity [20,21,69,70,74]. Within the laboratory environment, it has also been noted that clay mineralization induced by fungal pellets in suspension, may have diffusive effects on oxygen and nutrient uptake-exchange within the pellets, and that the clay particles can act as a barrier to remove toxic compounds [20].

3.2. Clays as Biosignatures

Iron is an omnipresent component of many, if not most, clays [75]. This is presumably a result of its great abundance in Earths' crust [75]. The valency and internal ratio of Fe^{2+} to Fe^{3+} within clay structures have a significant effect on the physiochemical properties of clays, such as their ability to expand, cation-exchange capacity (CEC) and surface area (e.g., [56,76]), meaning that variations in clay mineralogy may be linked to the redox capacity of the precipitating fluids. Because of the element-complexing properties found within the clay structure, clay minerals are known to form complexes with many different chemical species, including organic molecules [77–79]. This may be partly attributed to the interconnection between clays and iron, including Fe oxides, which has a special affinity to complex organic molecules that prevent the prevalent oxidation of organics [80,81]. Therefore, the suggested stepwise accretion of Fe onto microbial cells followed by subsequent attraction and complexation of Al and Si [7] may be responsible for the high organic content of clays. Since clays thus can act as a protective barrier of organics, this may be a relevant reason to focus on iron rich clays or reactive Fe-species when it comes to the search for biomarkers in sedimentary deposits on Earth or possibly on Mars [81,82].

A majority of studies concerning microbial clay authigenesis have focused on prokaryote communities in more or less shallow sedimentary environments. There is, however, a glaring gap in the knowledge of microbial clay authigenesis within the igneous crust of continents and ocean, which together represents some of the world's largest microbial habitats [83]. The oceanic crust can also provide an analogous setting for a potential subsurface-biosphere on Mars and understanding microbial preservation in cryptic subseafloor habitats on Earth may thus help us to better recognize similar processes on other planets. Therefore, the following sections will more specifically focus on mineralization of clays within the subsurface igneous crust, primarily but not exclusively, among fossil fungi. We will try to summarize the type of clay minerals that are primarily associated with biofilms in these types of extreme settings.

4. Microbial Clay Mineralization in the Igneous Crust: 2.4 Ga–48 Ma

4.1. Clay Fossilized Filaments from the Ongeluk Formation, South Africa

Ophiolitic pillow basalts from the Paleoproterozoic (2.4. Ga) Ongeluk formation, Griquatown West Basin, South Africa, was investigated by Bengtson et al. [19], who described filamentous fossils from secondary infilled amygdales in the basalt (Figure 1A–D). The Ongeluk basalt is estimated to have undergone low grade-type metamorphism, and the chlorite-mineralized filaments show metamorphic temperatures of 179–260 °C [19]. Individual filaments vary in size between 2–12 μm, but each filament has a consistent width throughout the length (Figure 1D). The filaments form intricate mycelium-like networks with morphological characteristics that suggest a fungal-like affinity (Figure 1D), which makes the Ongeluk fossils the currently oldest known fungus-like organisms in the fossil record, with large implications for the divergence of early opisthokonts [19].

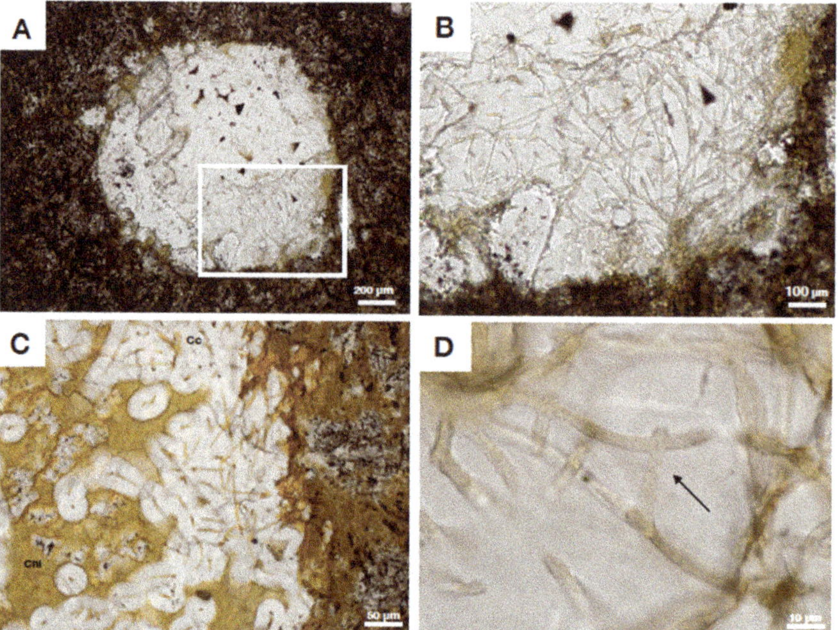

Figure 1. Optical micrographs of basalt from the Ongeluk formation, South Africa, showing a vesicle with extensive fungal-like mycelia consisting of chlorite. (**A**) Vesicle in basalt matrix with fungal chloritized hyphae surrounded by secondary infilled calcite. Boxed area is enlarged in (**B**). Scale bar equals 200 μm (B) Enlarged area from (A) showing dense fungal-like mycelia protruding from the walls into a vesicle. Filaments consist of chlorite and are enclosed in calcite. Scale bar is 100 μm. (**C**) Detail of chloritic fungal hyphae surrounded by calcite. Chlorite is yellow, calcite white. Scale bar equals 50 μm. (**D**) Chlorite filaments with even diameters show fungal like characteristics such as potential anastomosis (arrow). Scale bar equals 10 μm.

SEM-EDS analysis from the study showed that the chloritized filaments have elemental compositions corresponding to the basal fossil biofilms, from which they extend [19], suggesting a similar biological origin of both filaments and biofilm. All examined samples have matrices composed of chlorite, quartz, feldspar and calcite, with apatite and Fe-Ti oxides as main accessory minerals [19].

The metamorphic nature of the chloritized filaments made it difficult to identify the composition of the original clay phases of the Ongeluk filaments; a potential precursor mineral might have been a smectite such as saponite, or berthierine from the chlorine-group [84]. A secondary chlorite phase in the Ongeluk samples is associated with chalcopyrite and were thus presumably affected by a higher-degree metamorphism with subsequent circulation of hydrothermal fluids (i.e., [19]). Due to the endolithic characteristics of the fossils, the microorganisms must have entered the basalt during the time-window where the cooling cracks and vug-system of the basalt was still open to sea-water circulation [19], but prior to the closure of the system by secondary mineralizations. A conservative age estimate of the filaments was therefore 2.06 Ga; after the termination of fluid circulation within the rocks and closing of the amygdales [19]. By 2.06 Ga, the deep oceans were presumably largely anoxic [85], which would imply that the Ongeluk biota had a largely anaerobic lifestyle.

4.2. Fungal Fossils from Fractured Granitic Rock

Drake et al. [25] described the first known occurrence of presumed anaerobic fossil fungi from a cored borehole within fractured granitic rock from the Laxemar-site, near the Äspö research laboratory, in Sweden. The samples described in the study are from a sectioned core taken at 740 m depth, where fossil filamentous microbes show distinctive mycelia-like characteristics (Figure 2A–F). Due to difficulties of constraining the timing of colonization, as well as the age of secondary mineralizations, the exact age of the filaments remains unknown. They are however presumably of Phanerozoic age, resting in 1.8 Ga granite host rock. The filaments are preserved as partly kerogenous, and partly clay mineralized fossils (e.g., Figure 2C) and occur in open fractures that run through a quartz-vein [25]. Stable carbon isotope SIMS microanalyses showed that the calcite precipitated along the edges of the fossil-bearing crack have substantially negative isotope values, down to δ^{13}C-43‰ V-PDB, indicative of anaerobic methane oxidizing metabolisms, in association to the fungal communities [25]. The fungi have diameters ranging from 2–20 µm, and both EDS and Raman spectroscopic analysis reveal that the filaments consist of clay-like Fe-Mg-Ca- phases, as well as some minor Fe oxides [25].

For this study, we looked closer at two cores from the Laxemar area: XLX09 taken at 740 m depth and XCX04 taken at 678 m depth, to identify clay mineralogy and to characterize the redox state of iron by investigating the ratio of structural Fe^{2+}/Fe_{tot} within the clays (Figure 2D,E,G). The clay phase in the XLX09 core is a light beige-white substance, constricted to filamentous fungal hyphae (Figure 2D,E). The second core, XCX04, contains a more mature biofilm with less well-preserved hyphae and a darker, brown, clay phase (Figure 2G). Subsequent XRD analysis show that the main clay phase associated with the fungi from core XLX09, is a swelling smectite of dioctahedral montmorillonite-type (with some peaks also matching that of the trioctahedral smectite saponite), with a Fe^{2+}/Fe_{tot} ratio of 33.9 % (Figure 3A,B). The second core (XCX04), with a darker clay phase, showed no significant swelling during a second XRD run with prepared wet-samples, however the analysed peaks were closest matched by montmorillonite, again with some peak overlap of saponite. Fe^{2+}/Fe_{tot} was 30.6 %, i.e., it showed a small but significantly lower proportion of Fe^{2+} within the mineral structure compared to the white fungal clay phase in XLX09 (Figure 3A,C).

Figure 2. Scanning electron (**A–C,F,G**) and stereo (**D,E**) micrographs of fungal fossils from fractured granitic rock, Laxemar drill-cores, Sweden. The fossils are preserved mainly by clay mineralization but also as carbonaceous filaments. (**A**) Clay mineralized hyphae with a central strand. Scale bar equals 25 μm. (**B**) Close-up of clay mineralized hyphae with a central strand (arrow). Scale bar equals 25 μm. (**C**) Part clay-part carbonaceous fungal hyphae preserved within a hollow fracture. Scale bar equals 20 μm. (**D**) Sample of flight fungal biofilm from Laxemar run by XRD and Mössbauer spectroscopy for this review. Calcite surround the biofilm. Scale bar equals 300 μm. (**E**) Close-up of light fungal clay mineralized hyphae run by XRD and Mössbauer spectroscopy for this review. Scale bar equals 100 μm. (**F**) Poorly preserved fungal hyphae with uneven clay encrustation. Scale bar equals 300 μm. (**G**) XRD and Mössbauer spectroscopy sample showing a dark clay phase assumed to represent altered fungal biofilm from Laxemar. Scale bar equals 200 μm.

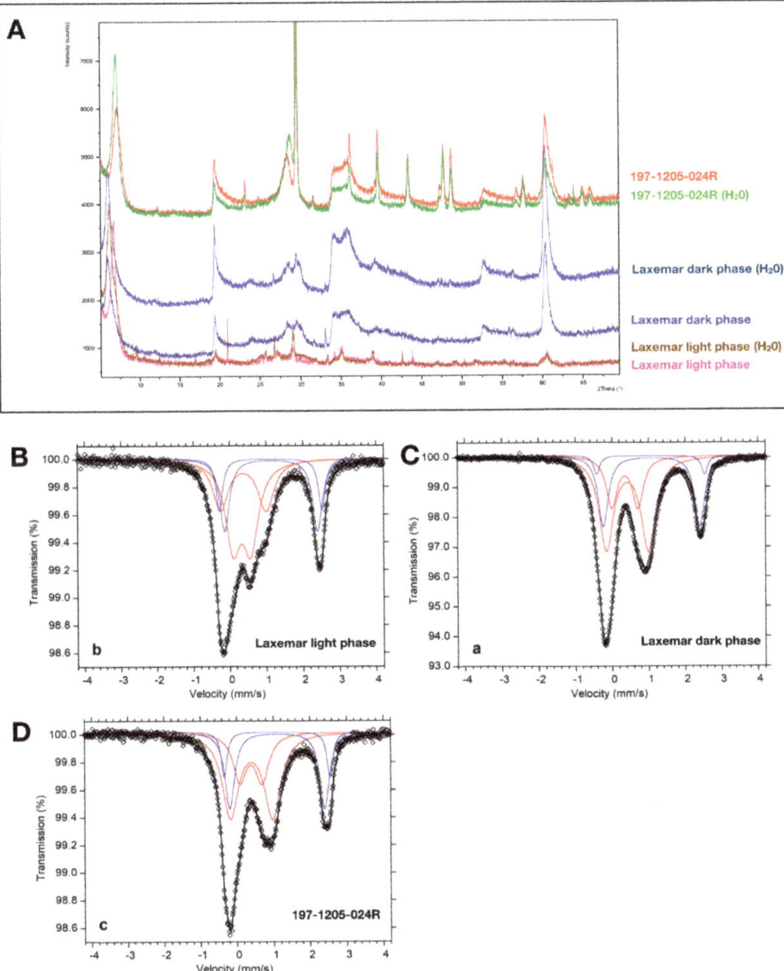

Figure 3. XRD (**A**) and Mössbauer spectra (**B–D**) of sample 1205A-024R (Nintoku seamount, clay mineralized hyphae) and two fungal biofilm samples (light and dark) from Laxemar, Sweden. (**A**) Summary of all XRD spectra encoded by color showing XRD peaks most closely related to smectites, in particular montmorillonite (there are also some peak matches with the smectite-group mineral saponite). (**B**) Mössbauer spectrum of light-colored clay mineralized fungal filaments from Laxemar. Diamonds represent the measured spectrum; the thick solid line represents the sum of the two fitted doublets (thin lines) assigned to Fe^{3+} and Fe^{2+} (Fe^{2+}/Fe_{tot}: 33.9%). (**C**) Mössbauer spectrum of slightly altered dark-colored clay mineralized biofilm from Laxemar. Diamonds represent the measured spectrum; the thick solid line represents the sum of the two fitted doublets (thin lines) assigned to Fe^{3+} and Fe^{2+} (Fe^{2+}/Fe_{tot}: 30.6%). (**D**) Mössbauer spectrum of green-colored clay mineralized fungal filaments from sample 1205A-024R, Nontoku seamount. Diamonds represent the measured spectrum; the thick solid line represents the sum of the two fitted doublets (thin lines) assigned to Fe^{3+} and Fe^{2+} (Fe^{2+}/Fe_{tot}: 33.5%).

4.3. Fungal Biofilms within The Emperor Seamounts

Together with Hawaii, the Emperor seamounts, a submarine volcanic chain named in large after famous Japanese Emperors, extends over 5000 km in the Pacific Ocean and contains a sequence of volcanic islands and seamounts, presumably resulting from hotspot volcanism [25]. The chain has a north-south trend after which it bends to the south east- the oldest part, approximately 81 Ma, lies in the North East and the chain becomes successively younger until, by about 43 Ma, it bends towards the younger Hawaii islands.

Detroit seamount is situated in the northernmost, and oldest, part of the Emperor chain, and was drilled at several locations in 2001 (ODP leg197, cores 1203, 1204A and 1204B) [86]. The seamount holds pillow basalts with an age of approximately 81 Ma [35], dating back to the Cretaceous period. Within these cores, Ivarsson et al. [15,16], reported the presence of clay-fossilized fungi from a depth of 936.65 mbsf within the seamount (Figure 4A–D). The fossil fungal filaments and sporophore-like structures (e.g., Figure 4D) were closely associated with the presence of botryoidal Mn-oxides (Figure 4A–C), suggested to have a biological and possibly fungal, origin [16]. The fossil biofilms, green in stereo microscopic-reflected light, were identified in open vesicles (Figure 4A,C) and the filaments are attached to a basal biofilm, from which they protruded into the open space of the vug [16].

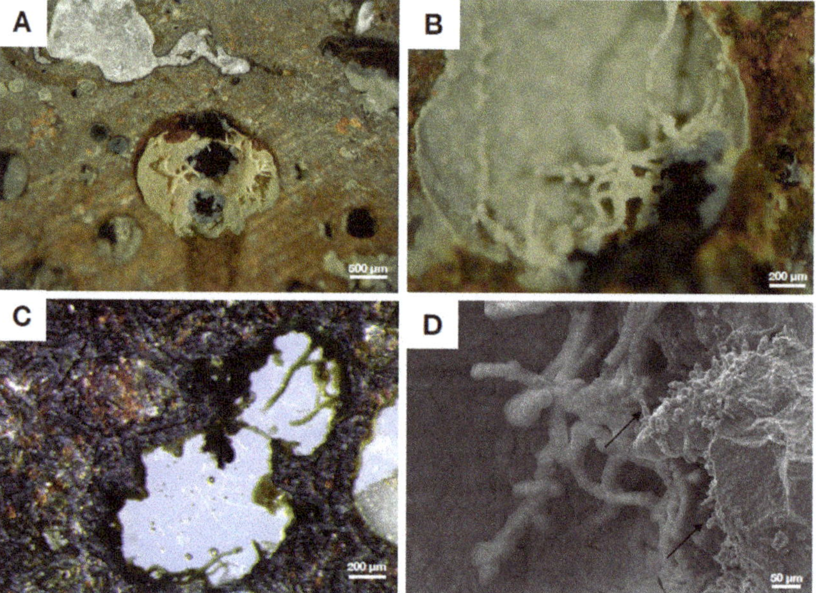

Figure 4. Stereo (**A–C**) and Scanning Electron micrographs of fungal biofilms protruding from the walls into empty or partly filled vescicles within basalt from Detroit seamount, samples 1204B-16R-01 and 1203A-57R3. (**A**) vesicle filled with authigenic clay (yellow-cream) and Mn-oxides (black). Fungal hyphae are preserved in clay within the vesicle. Scale bar equals 500 μm. (**B**) Close-up of hyphae and basal biofilm preserved as clay with a basalt matrix. Scale bar equals 200 μm. (**C**) Vesicle with green-colored clay mineralized filaments and biofilm protruding into an empty void surrounded by basalt matrix. Scale bar equals 200 μm. (**D**) Fungal mycelia with potential sporophore structures (arrows). Scale bar equals 50 μm.

The basal biofilm as well as the filaments have similar compositions and consist of Fe-rich smectites [16]. The smectite phases were analysed using Raman spectroscopy, and showed distinct peaks most closely corresponding to Nontronite, the Fe^{3+}-rich smectite endmember. A few spectral variations however suggested the possibility of a mixed signal for the clay phases, possibly corresponding to a Ca-poor montmorillonite (e.g., see [16]).

Ivarsson et al. [17], also investigated basaltic drill core samples from the Nintoku seamount, situated centrally within the Emperor submarine chain of Paleocene/Eocene age, dated to approximately 56 Ma.

The authors examined a cored section from IODP sample site 197-1206-34R, and therein described an extensive fracture system, split open to expose fossil biofilms and associated microbial structures which had formed along the fracture-walls (Figure 5A–F). The exposed biofilm was 20–100 µm thick and contained cell-like structures comparable to yeast growth-phases (Figure 5E,F). Filamentous fungi, 15–25 µm in diameter, extended from the basal film in a complex mycelia-like fashion, comparable to that of other samples from the Emperor chain (e.g., Ivarsson et al. [15,16]. XRD and Raman spectroscopy analysis showed that both biofilm and filaments consisted of a swelling smectite-layer, most closely related to montmorillonite. In association to the hypha, cauliflower-like hematite bodies similar to microstromatolites were found, consisting of banded hematite [17].

Another IODP sample (197-1205A-024R) from Nintoku seamount was analysed using XRD and Mössbauer spectroscopy for this review in order to ascertain the clay composition and iron redox state at a different location within the seamount. The sample contained a green clay phase with a basal biofilm and filaments corresponding in morphology to fungal hyphae similar to those described by Ivarsson et al. above (Figure 5G). The hyphae consisted of a swelling smectite phase closely corresponding to XRD spectra from montmorillonite (with some peaks also matching that of saponite), with Fe^{2+}/Fe_{tot} ratio of 33.5 % (Figure 3A,D). These results correspond well to the XRD results obtained by Ivarsson et al. [17] from other depths at site 1205A and suggests a consistency with respect to the main clay phases present at different locations and depths within the Nintoku seamount, but also between different seamounts within the Emperor chain as well.

One of the youngest seamounts in the Emperor chain is Koko seamount, an underwater volcano dated to the Eocene epoch at 48 Ma [35]. Koko seamount rises approximately 5000 m from the abyssal plain, and Bengtson et al. [18] described from IODP sample 197-1206A-4R filamentous mycelia-like networks consisting of hematite tubules surrounded by a montmorillonite-like clay layer that extended from partly open vugs and vesicles in the basalt (Figure 6A–D). The fossil filaments were attached to a basal biofilm, with an outer crust consisting of a montmorillonite-phase similar to those surrounding the filaments. The base of the biofilm, however, was found to consist of hematite with carbon, and an upper hematite layer and montmorillonite type clay on top [18]. The hyphae extending from the basal film were often associated with cauliflower-like iron rich microstromatolites (Figure 6D), similar to the fungal systems present in Nintoku seamount (e.g., [17]).

Figure 5. Stereo (**A–D,G**) and Scanning Electron micrographs showing fungal biofilm with yeast-like structures preserved in green clay from samples 1205-34R-5 (**A–F**) and 1206A-024R (**G**), Nintoku seamount. (**A**) Open fracture exposing filamentous fungal biofilms preserved in clay, partly intergrown and overgrown by zeolites (translucent minerals) with dark Mn-oxides present. Boxed area is magnified in (**B**). Scale bar is 1 mm. (**B**) Close-up of boxed area from (**A**) showing clay mineralixed hyphae intergrown with zeolite crystal. Boxed area is magnified in (**C**). Scale bar is 200 µm. (**C**) Close-up of boxed area from (**B**). Evenly preserved clay mineralized fungal hyphae with spore-like structures attached. Scale bar is 50 µm. (**D**) Fungal mycelium intergrown with zeolites. Scale bar is 100 µm. (**E**) Fungal biofilm with yeast-like cells attached to the surface. Scale bar is 250 µm. (**F**) Close-up of yeast-like cells from fungal biofilm. All cells and hyphae are preserved as authigenic clays. Scale bar is 50 µm. (**G**) XRD and Mössbauer-run sample 1206A-024R, showing green clay mineralized fungal filaments. Scale bar is 100 µm.

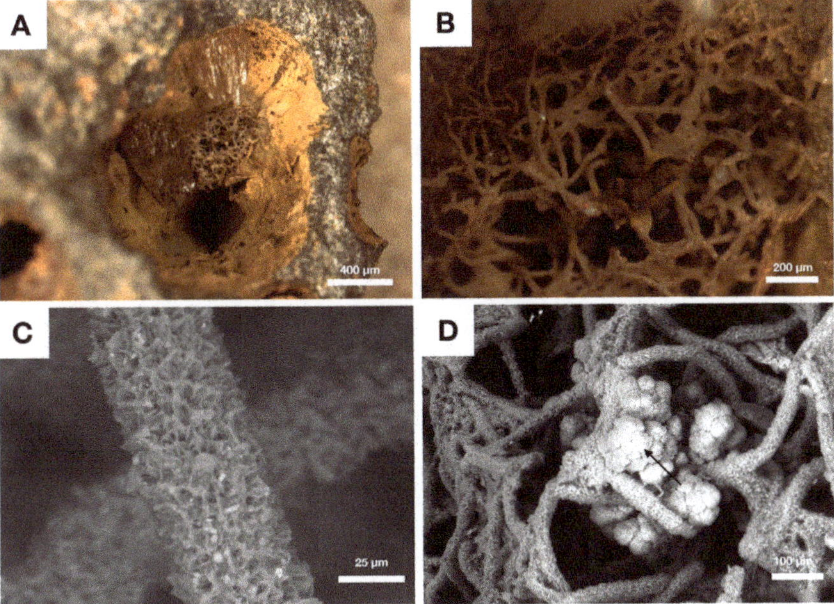

Figure 6. Stereo (**A–B**) and Scanning Electron micrographs (**C–D**) of clay mineralized fungal biofilms from Koko seamount, sample 1206-4R-2. (**A**) Vesicle containing clay encrusted mycelia that branch and diverge. Scale bar is 400 µm. (**B**) Close-up of pristine mycelia network preserved in a rust-colored clay. Scale bar is 200 µm. (**C**) Close-up of evenly clay encrusted fungal hyphae. Scale bar is 24 µm. (**D**) Fungal mycelia with embedded cauliflower-like microstromatolites of *Frutexites*-type (arrow).

4.4. Authigenic Clays within Late Devonian Pillow Basalts

Eickmann et al. [14] described the presence of fossilized cryptoendolithic filaments, of uncertain biological affinity, found in calcite-filled vesicles from ophiolitic Late Devonian pillow basalts in Germany. The filaments were preserved by clay mineralization, with an endmember composition similar to berthierine (with 36% of Fe as Fe^{2+})-chamosite/illite for the central strand, and illite-glauconite precipitated like a halo around the central strand. The precipitation of clay as described by Eickmann et al. [14] presumably proceeded as a passive mineralization of active functional groups on filament surfaces, similar to the process described by [7]. The most pristine filaments in the study contained Chamosite, the Fe^{2+} endmember of the chlorite group, indicating that encrustation of microbial cells proceeded from a poorly ordered Fe, Al silicate phase to a more crystalline chamosite (e.g., [2,13]). Similarly, Peckmann et al. [13] described microbial clay encrustation analogous to the Fe-Si encrustation explained from many other studies of mainly prokaryote biofilms (e.g., [1–5,7] and references therein), related to cryptoendolithic filamentous microorganisms from another German ophiolitic pillow basalt. Two kind of filaments were identified in the study; the first with an illite center consisting of a thin outer chamosite rim, and the other with a central TiO_2 strand surrounded by illite with an outer rim of chamosite.

5. Redox Variability in Crustal Biosphere Habitats

Studies of microbial clay authigenesis in the igneous crust have so far only scratched the surface. However, the antiquity of for example the fungus-like Ongeluk fossils (Figure 1) described by Bengtson et al. [17], testifies to the fundamental nature of microbial clay mineralization in the deep subseafloor crust, extending as far back as the Paleoproterozoic. It is apparent from many studies i.e., [1–5,7–10,13–19,25,33–39] that authigenic mineral precipitation, in particular clay authigenesis, related to microbial communities and their EPS seems to follow a similar pattern (see e.g., [7]) among prokaryotes as well as within most fungal biofilms. This suggests an analogous mode of formation, albeit with some variations, particularly in the style and degree of microbial preservation within the igneous crust. Variations in the orogenic history of the source rock, may be one reason for these differences. How variations in oxygen levels and thus the redox state of a system affects fungal clay authigenesis, including the type of clay that forms, is however more uncertain: Seamounts and oceanic spreading centers are commonly uncovered by sediments, and can work as oxygen pumps of sorts, circulating oxygenated seawater through the porous upper oceanic crust, i.e., [87]. Generally, within the deep ocean sediments, oxygen is rapidly used up through reactions with organic material in the shallow upper layers, which leave significant parts of deep-sea sediments essentially anoxic [87]. At the sediment-basalt interface, however, increased oxygen concentrations are to be expected due to the input of oxygen-rich oceanic water circulating within the crust and diffusing upwards, into the sediments [87]. The description of especially Fe and Mn oxides in amygdales and cracks within oceanic basalts (e.g., [16,35]) are good indications of at least partly oxygenated conditions within the oceanic crust, permitting the transportation of otherwise insoluble Fe and Mn species. However, the oceanic crust is heterogenous, and even if the oceanic crust in general is believed to be more or less oxygenated [88], locally reducing conditions may prevail in places where anoxic or sub-oxic fluids from hydrothermal sources reach the crust [87] or in areas extending away from mid-ocean ridges and hotspots.

Oxygenation within a sealed space may also be subject to variations on the temporal scale during which colonization of the basalt takes place affecting the subsequent precipitation of secondary minerals: For example, McKinley et al. [11], describe clay authigenesis in Miocene flood basalts, where fractures containing clay-fossilized filaments of presumed prokaryotic nature also contain a range of secondary minerals such as oxic ferrihydrite, trioctahedral Fe^{2+}-rich smectites, pyrite and quartz, together suggesting that local conditions were initially oxic, but turned progressively anoxic [11], emphasizing the variable nature of the basaltic crust.

Our review of authigenic clay precipitation within fungal communities suggest that the impact of oxygen on the precipitating clay phase may in fact be rather negligible, at least within the deep crust, seeing as the preferred mineralogy associated with fossil fungi from both anoxic deep granite rocks as well as oxygenic seamount basalts, are similar types of swelling montmorillonites (see Table 1 for comparison of clay mineralization among various herein discussed localities). Similarities in our XRD spectra with that of the trioctahedral smectite saponite is also worth to mention, seeing as saponite clay have been attributed to the activity of microbial communities in hydrothermal settings from Iceland, recently described by Geptner et al. [53]. Various members of the chlorite group have been more sporadically described associated with deep biosphere microbial fossils, i.e., [13,14,19], all with ages ranging from Devonian or older. Also, the degree of structural Fe^{2+} is roughly equal between these settings, with a small but significant difference of about 2.9% more Fe^{2+} within montmorillonites/smectites precipitated under anoxic conditions.

Table 1. Summary of studies focusing on subsurface clay authigenesis associated with (mainly) fungal biofilms. Publications are arranged according to age from the top-down.

Reference	Age	Setting	Preservation	Clay Phase	Color of Clay	Assumed Source Fluid (This Review)
[19]	Bedrock 2.4 Ga (filament ages ~2.06 Ga)	Ongeluk ophiolite, South Africa	Clay mineralization, slightly metamorphic and chloritized	Chlorite (transformed from precursor clay)	Pale yellow-brown	Anoxic
[25]	Bedrock 1.8 Ga (filament ages not constrained, Phanerozoic)	Continental fractured granite Laxemar, Sweden (drill cores)	Clay mineralization/carbonaceous filaments	Fe/Mg/Ca-rich clay, minor Fe-oxides	Cream white (sample XLX09)	Anoxic
[13]	Presumed Devonian	Arnstein ophiolite, Germany (pillow basalt)	Clay mineralization	TiO_2 filament-strands surrounded by illite and chamosite	Pale yellow-brown	Presumed oxic-sub oxic
[14]	Presumed Devonian	Frankenwald and Thüringer Wald, Germany (pillow basalt)	Clay mineralization	Illite center/chamosite rim or TiO_2 filament-strands surrounded by illite and chamosite	Green	Presumed oxic-sub oxic
[16]	Cretaceous, 81 Ma	Oceanic crust, Detroit seamount (basalt drill cores)	Clay mineralization	Smectite (montmorillonite)	Cream yellow (sample 197-1204B-16R-01)	Oxic -slightly sub oxic
[17]	Paleocene-Eocene boundary, 56 Ma	Oceanic crust, Nintoku seamount (basalt drill cores)	Clay mineralization	Smectite (montmorillonite)	Green (sample 197-1205-34R-5)	Oxic -slightly sub oxic
[15]	Paleocene-Eocene boundary, 56 Ma (Nintoku seamount)/ Eocene, 48 Ma (Koko seamount)	Oceanic crust, Nintoku and Koko seamounts (basalt drill cores)	Clay mineralization	Smectite (montmorillonite)	Green (sample 197-1205-34R-5), Yellow-cream/red (sample 197-1206-4R-2)	Oxic -slightly sub oxic
[18]	Eocene, 48 Ma	Oceanic crust, Koko seamount (basalt drill cores)	Clay mineralization	Smectite (montmorillonite)	Yellow-cream/red (sample 197-1206-4R-2)	Oxic -slightly sub oxic

Noteworthy is that the preservation of filamentous fungi from the Pacific Ocean Emperor seamounts appears to have occurred in vivo, or possibly very early *post mortem*, considering the pristine nature of the clay-encrusted filaments (i.e., Figures 4–6). The mycelia are evenly encrusted and structurally intact, forming complex 3D networks (i.e., Figure 6). They do not appear collapsed, and filaments can rather often be seen protruding in or out of associated secondary minerals such as zeolites (Figure 5A–D). All filaments described by Ivarsson et al. and Bengtson et al. from the Emperor seamounts [15–19,33–38], are thus encrusted by mainly smectites in the form of montmorillonite-type clays. Swelling smectites such as montmorillonite in particular, are particularly sensitive to alteration processes that can affect the structure and mineralogy of the clay, i.e., [89]. Smectites are also among the most prone to destabilization due to for example microbial iron reduction [90], which make smectites including montmorillonite-type clays good indicators of fairly juvenile, or unaltered, conditions within sealed vesicles or cracks. These characteristics of smectites corresponds well to the structural fidelity of the montmorillonite-encrusted fungi from especially the Emperor seamounts. Intriguingly, the fungi preserved by clay mineralization in the anoxic deep crust, exemplified here by the Laxemar fungal fossils (Figure 2), appear less well-preserved, compared with the pristine fossils of basaltic seamounts (i.e., compare Figure 2C,F to Figures 5 and 6). In a few cases, it is possible to observe that the Laxemar filaments are only partly encrusted with montmorillonite clays and the remainder of the filament is instead preserved as kerogenous fossils (Figure 2C). In all, mineralization in these settings appear to have been less continuous, and this together with the irregularity of preserved specimens, might suggest that mineralization did not occur while the organisms were alive, but rather as a post mortem process on top of carbonaceous filaments. Thus, the presence of oxygen in the ambient environment appears to have a small effect on the type of clay mineral that forms. However, the effect of oxygen may be larger when it comes to the timing and mode of fossil preservation, especially related to the preservation of kerogenous filaments, which may be more pronounced in low oxygen settings.

One potential driving mechanism behind the structural differences exhibited by montmorillonite-covered fungi in anoxic granitic versus oxic basaltic settings, may be related to the production and structure of fungal EPS, which can differ greatly between oxygen-variable settings; In general, fungal EPS production is stimulated by the presence of oxygen and can be lower or absent among anaerobic fungi [62]. Fungal EPS, like microbial EPS in general, is expected to play a significant part in the mineralization process, acting as a cation-trap and potential mineral nucleation-source, eventually promoting further mineral precipitation [45]. Therefore, the amount as well as the structure of the fungal EPS may well affect the degree and style of mineralization within fungal communities as well.

Looking closer at some of the ophiolitic settings described by Bengtson et al. [19], Peckmann et al. [13] and Eickmann et al. [14], the taphonomy of microbial preservation here seems rather similar; with filaments preserved as a central strand surrounded first by a halo and then by an outer rim, with different mineral compositions. This may suggest that various physical and diagenetic processes related to the uplift of oceanic crust onto continents have affected both the mineralogy and taphonomy of the cryptoendolithic microbes. The abundance of chlorite mineral members from all of these locations support such a theory, seeing as chlorite derived from various precursor clay minerals are common processes during low temperature (<200 °C) alterations (e.g., [84]). The complex nature of preservation, with several different mineral phases or "zones", may thus be the result of a complex diagenetic and orogenic history, compared to microbial fossils containing only one juvenile clay phase, such as smectites of montmorillonite-type (see e.g., [89]).

Further supporting a more complex mode of microfossil-preservation in relation to low temperature alterations, are microfossil-finds by Sakakibara et al. [91] within pre-Jurassic age metabasaltic rocks from central Shikoku, Japan. This study described the presence of branching filamentous microbes within previously water-filled vesicles within the basalt, which allowed the formation of microbial clay within the open space, thus preserving the putative microorganisms. In the case of the Sakakibara [91] fossils, these are preserved as central filamentous strands composed of Fe oxides and surrounded by a halo of phengite with a rim of pumpellyite—all of which could have formed from microbial clay precursors such as illite-glauconite (for phengite) and illite-chamosite (for pumpellyite) during light metamorphism [91].

6. Summary

Preservation of cryptoendolithic microbial communities, especially fungi, within Earths igneous crust is intimately linked to the precipitation of authigenic clays. In particular swelling smectite minerals such as montmorillonite, have been described from various deep subsurface localities. Our study suggests a larger input from biology on the formation of clays in the subsurface igneous crust of both land and oceans, than what has hitherto been recognized. While traditionally, the external environment has been suggested to control the mineralogy of precipitating clays to a large extent, our results suggest that similar type of swelling smectites form associated with fungal filaments and biofilms in subsurface crustal settings highly variable with respect to both source rock (felsic granite versus basic basalt) and ambient pO_2. The effect of oxygen on fungal clay precipitation, however, is likely complex, and may affect the timing as well as the taphonomy of the mineralized microbe. Overall, we suggest that intrinsic factors, such as the extracellular organic matrix, have a large impact on the type of clay that precipitates around microbial cells and EPS. This, in turn, suggests that the presence of life in the deep oligotrophic crust probably have a far greater effect on the cycling of chemical species, such as Fe, Al and Si than what has been previously established and may in fact be responsible for a great fraction of secondary clay minerals in the deep subsurface biosphere.

Author Contributions: T.S. designed and implemented the study, wrote the initial manuscript and aided with sample analysis. M.I. contributed with manuscript conception and providing samples. H.D. contributed with manuscript conception, C isotope data used in the review and providing samples. H.S. contributed with XRD, Mössbauer analysis and manuscript conception.

Funding: This research was funded by a Carl Trygger Foundation post-doctoral grant (CTS 18:167) to TS and MI, a Swedish Research Council grant (contracts No. 2012-4364, 2017-04129) to MI, a Villum Investigator grant to Donald Canfield (No. 16518), a Swedish Research Council grant (contract 2017-05186) and a Formas grant (contract 2017-00766) to HD.

Acknowledgments: The authors wish to thank the Swedish Nuclear Fuel and Waste Management Co (SKB) for access to drill core samples.

Conflicts of Interest: The authors declare no conflict of interest.

References

1. Ferris, F.G.; Beveridge, T.J.; Fyfe, W.S. Iron-silica crystallite nucleation by bacteria in a geothermal sediment. *Nature* **1986**, *320*, 609–611. [CrossRef]
2. Ferris, F.; Fyfe, W.; Beveridge, T. Bacteria as nucleation sites for authigenic minerals in a metal-contaminated lake sediment. *Chem. Geol.* **1987**, *63*, 225–232. [CrossRef]
3. Ferris, F.G.; Fyfe, W.S.; Beveridge, T.J. Bacteria as nucleation sites for authigenic minerals. In *Developments in Geochemistry*; Elsevier: Amsterdam, The Netherlands, 1991; Volume 6, pp. 319–325.
4. Konhauser, K.O.; Fyfe, W.S.; Ferris, F.G.; Beveridge, T.J. Metal sorption and mineral precipitation by bacteria in two Amazonian river systems: Rio Solimões and Rio Negro, Brazil. *Geology* **1993**, *21*, 1103–1106. [CrossRef]
5. Konhauser, K.O.; Schultze-Lam, S.; Ferris, F.G.; Fyfe, W.S.; Longstaffe, F.J.; Beveridge, T.J. Mineral Precipitation by Epilithic Biofilms in the Speed River, Ontario, Canada. *Appl. Environ. Microbiol.* **1994**, *60*, 549–553. [PubMed]
6. Chafetz, H.S.; Buczynski, C. Bacterially induced lithification of microbial mats. *Palaios* **1992**, *7*, 277–293. [CrossRef]
7. Konhauser, K.O.; Urrutia, M.M. Bacterial clay authigenesis: a common biogeochemical process. *Chem. Geol.* **1999**, *161*, 399–413. [CrossRef]
8. Knorre, H.V.; Krumbein, W.E. Bacterial calcification. In *Microbial Sediments*; Springer: Berlin/Heidelberg, Germany, 2000; pp. 25–31.

9.   Konhauser, K.O.; Hamade, T.; Raiswell, R.; Morris, R.C.; Ferris, F.G.; Southam, G.; Canfield, D.E. Could bacteria have formed the Precambrian banded iron formations? *Geology* **2002**, *30*, 1079–1082. [CrossRef]

10.   Vasconcelos, C.; Warthmann, R.; McKenzie, J.A.; Visscher, P.T.; Bittermann, A.G.; van Lith, Y. Lithifying microbial mats in Lagoa Vermelha, Brazil: modern Precambrian relics? *Sediment. Geol.* **2006**, *185*, 175–183. [CrossRef]

11.   McKinley, J.P.; Stevens, T.O.; Westall, F. Microfossils and paleoenvironments in deep subsurface basalt samples. *Geomicrobiol. J.* **2000**, *17*, 43–54.

12.   Schiffman, P.; Fisher, Q.J.; Konhauser, K.O. Microbial mediation of authigenic clays during hydrothermal alteration of basaltic tephra, Kilauea Volcano. *Geochem. Geophys. Geosystems* **2002**, *3*, 1–13.

13.   Peckmann, J.; Bach, W.; Behrens, K.; Reitner, J. Putative cryptoendolithic life in Devonian pillow basalt, Rheinisches Schiefergebirge, Germany. *Geobiology* **2008**, *6*, 125–135. [CrossRef]

14.   Eickmann, B.; Bach, W.; Kiel, S.; Reitner, J.; Peckmann, J. Evidence for cryptoendolithic life in Devonian pillow basalts of Variscan orogens, Germany. *Palaeogeogr. Palaeoclim. Palaeoecol.* **2009**, *283*, 120–125. [CrossRef]

15.   Ivarsson, M.; Bengtson, S.; Skogby, H.; Belivanova, V.; Marone, F. Fungal colonies in open fractures of subseafloor basalt. *Geo-Marine Lett.* **2013**, *33*, 233–243. [CrossRef]

16.   Ivarsson, M.; Broman, C.; Gustafsson, H.; Holm, N.G. Biogenic Mn-oxides in subseafloor basalts. *PLoS ONE* **2015**, *10*. [CrossRef]

17.   Ivarsson, M.; Bengtson, S.; Skogby, H.; Lazor, P.; Broman, C.; Belivanova, V.; Marone, F. A fungal-prokaryotic consortium at the basalt-zeolite interface in subseafloor igneous crust. *PLoS ONE* **2015**, *10*. [CrossRef]

18.   Bengtson, S.; Ivarsson, M.; Astolfo, A.; Belivanova, V.; Broman, C.; Marone, F.; Stampanoni, M. Deep-biosphere consortium of fungi and prokaryotes in Eocene subseafloor basalts. *Geobiology* **2014**, *12*, 489–496. [CrossRef]

19.   Bengtson, S.; Rasmussen, B.; Ivarsson, M.; Muhling, J.; Broman, C.; Marone, F.; Stampanoni, M.; Bekker, A. Fungus-like mycelial fossils in 2.4-billion-year-old vesicular basalt. *Nat. Ecol. Evol.* **2017**, *1*, 141. [CrossRef]

20.   Fomina, M.; Gadd, G.M. Influence of clay minerals on the morphology of fungal pellets. *Mycol. Res.* **2002**, *106*, 107–117. [CrossRef]

21.   Fomina, M.; Gadd, G.M. Metal sorption by biomass of melanin-producing fungi grown in clay containing medium. *J. Chem. Technol. Biotechnol.* **2003**, *78*, 23–34. [CrossRef]

22.   Konhauser, K.O. Diversity of bacterial iron mineralization. *Earth-Science Rev.* **1998**, *43*, 91–121. [CrossRef]

23.   Konhauser, K.O.; Ferris, F.G. Diversity of iron and silica precipitation by microbial mats in hydrothermal waters, Iceland: Implications for Precambrian iron formations. *Geology* **1996**, *24*, 323. [CrossRef]

24.   Wierzchos, J.; Ascaso, C.; Sancho, L.G.; Green, A. Iron-Rich Diagenetic Minerals are Biomarkers of Microbial Activity in Antarctic Rocks. *Geomicrobiol. J.* **2003**, *20*, 15–24. [CrossRef]

25.   Drake, H.; Ivarsson, M.; Bengtson, S.; Heim, C.; Siljeström, S.; Whitehouse, M.J.; Broman, C.; Belivanova, V.; Åström, M.E. Anaerobic consortia of fungi and sulfate reducing bacteria in deep granite fractures. *Nat. Commun.* **2017**, *8*, 55. [CrossRef]

26.   Edwards, K.J.; Rogers, D.R.; Wirsen, C.O.; McCollom, T.M. Isolation and characterization of novel psychrophilic, neutrophilic, fe-oxidizing, chemolithoautotrophic α- and γ-Proteobacteria from the Deep Sea. *Appl. Environ. Microbiol.* **2003**, *69*, 2906–2913. [CrossRef]

27.   Edwards, K.J.; Fisher, A.T.; Wheat, C.G. The Deep Subsurface Biosphere in Igneous Ocean Crust: Frontier Habitats for Microbiological Exploration. *Front. Microbiol.* **2012**, *3*, 8. [CrossRef]

28.   Orcutt, B.N.; Sylvan, J.B.; Knab, N.J.; Edwards, K.J. Microbial ecology of the Dark Ocean above, at, and below the seafloor. *Microbiol. Mol. Boil. Rev.* **2011**, *75*, 361–422. [CrossRef]

29.   Thorseth, I.; Torsvik, T.; Furnes, H.; Muehlenbachs, K. Microbes play an important role in the alteration of oceanic crust. *Chem. Geol.* **1995**, *126*, 137–146. [CrossRef]

30.   Staudigel, H.; Furnes, H.; McLoughlin, N.; Banerjee, N.R.; Connell, L.B.; Templeton, A. 3.5 billion years of glass bioalteration: Volcanic rocks as a basis for microbial life? *Earth-Sci. Rev.* **2008**, *89*, 156–176. [CrossRef]

31.   Banerjee, N.R.; Izawa, M.R.M.; Sapers, H.M.; Whitehouse, M.J. Geochemical biosignatures preserved in microbially altered basaltic glass. *Surf. Interface Anal.* **2011**, *43*, 452–457. [CrossRef]

32.   Schumann, G.; Manz, W.; Reitner, J.; Lustrino, M. Ancient fungal life in North Pacific Eocene oceanic crust. *Geomicrobiol. J.* **2004**, *21*, 241–246. [CrossRef]

33.   Ivarsson, M.; Lindblom, S.; Broman, C.; Holm, N.G. Fossilized microorganisms associated with zeolite–carbonate interfaces in sub-seafloor hydrothermal environments. *Geobiology* **2008**, *6*, 155–170. [CrossRef]

34. Ivarsson, M.; Lausmaa, J.; Lindblom, S.; Broman, C.; Holm, N.G. Fossilized microorganisms from the Emperor Seamounts: Implications for the search for a subsurface fossil record on Earth and Mars. *Astrobiology* **2008**, *8*, 1139–1157. [CrossRef]

35. Ivarsson, M.; Bengtson, S.; Belivanova, V.; Stampanoni, M.; Marone, F.; Tehler, A. Fossilized fungi in subseafloor Eocene basalts. *Geology* **2012**, *40*, 163–166. [CrossRef]

36. Ivarsson, M.; Holm, N.G.; Neubeck, A. The deep biosphere of the subseafloor igneous crust. In *Trace Metal Biogeochemsitry and Ecology of Deep-Sea Hydrothermal Vent Systems*; Demina, L.L., Galkin, S.V., Eds.; Springer: Cham, Switzerland, 2015; pp. 1–24.

37. Ivarsson, M.; Schnürer, A.; Bengtson, S.; Neubeck, A. Anaerobic Fungi: A Potential Source of Biological H2 in the Oceanic Crust. *Front. Microbiol.* **2016**, *7*, 674. [CrossRef]

38. Ivarsson, M.; Bengtson, S.; Neubeck, A. The igneous oceanic crust – Earth's largest fungal habitat? *Fungal Ecol.* **2016**, *20*, 249–255. [CrossRef]

39. Drake, H.; Ivarsson, M.; Tillberg, M.; Whitehouse, M.; Kooijman, E. Ancient microbial activity in deep hydraulically conductive fracture zones within the forsmark target area for geological nuclear waste disposal, Sweden. *Geosciences* **2018**, *8*, 211. [CrossRef]

40. Drake, H.; Ivarsson, M. The role of anaerobic fungi in fundamental biogeochemical cycles in the deep biosphere. *Fungal Boil. Rev.* **2018**, *32*, 20–25. [CrossRef]

41. Sutherland, I.W. Biofilm exopolysaccharides: a strong and sticky framework. *Microbiol.* **2001**, *147*, 3–9. [CrossRef]

42. Flemming, H.C.; Neu, T.R.; Wozniak, D.J. The EPS matrix: the "house of biofilm cells". *J. Bacteriol.* **2007**, *189*, 7945–7947. [CrossRef]

43. Costerton, J.W.; Lewandowski, Z.; Caldwell, D.E.; Korber, D.R.; Lappin-Scott, H.M. Microbial biofilms. *Annu. Rev. Microbiol.* **1995**, *49*, 711–745. [CrossRef]

44. Arp, G.; Reimer, A.; Reitner, J. Calcification in cyanobacterial biofilms of alkaline salt lakes. *Eur. J. Phycol.* **1999**, *34*, 393–403. [CrossRef]

45. Arp, G. Photosynthesis-Induced Biofilm Calcification and Calcium Concentrations in Phanerozoic Oceans. *Science* **2001**, *292*, 1701–1704. [CrossRef]

46. Dupraz, C.; Visscher, P.T. Microbial lithification in marine stromatolites and hypersaline mats. *Trends Microbiol.* **2005**, *13*, 429–438. [CrossRef]

47. Dupraz, C.; Reid, R.P.; Braissant, O.; Decho, A.W.; Norman, R.S.; Visscher, P.T. Processes of carbonate precipitation in modern microbial mats. *Earth-Science Rev.* **2009**, *96*, 141–162. [CrossRef]

48. Hazen, R.M.; Schiffries, C.M. Why deep carbon? *Rev. Mineral. Geochem.* **2013**, *75*, 1–6. [CrossRef]

49. Prescher, C.; McCammon, C.; Dubrovinsky, L. MossA: A program for analyzing energy-domain Mössbauer spectra from conventional and synchrotron sources. *J. Appl. Crystallogr.* **2012**, *45*, 329–331. [CrossRef]

50. Fortin, D.; Ferris, F.G.; Scott, S.D. Formation of Fe-silicates and Fe-oxides on bacterial surfaces in samples collected near hydrothermal vents on the Southern Explorer Ridge in the Northeast Pacific Ocean. *Am. Miner.* **1998**, *83*, 1399–1408. [CrossRef]

51. Léveillé, R.J.; Datta, S. Lava tubes and basaltic caves as astrobiological targets on Earth and Mars: A review. *Planet. Space Sci.* **2010**, *58*, 592–598. [CrossRef]

52. Sánchez- Navas, A.S.; Algarra, A.M. Nieto Bacterially-mediated authigenesis of clays in phosphate stromatolites. *Sedimentolopgy* **1998**, *45*, 519–533. [CrossRef]

53. Geptner, A.; Kristmannsdottir, H.; Kristjansson, J.; Marteinsson, V. Biogenic Saponite from an Active Submarine Hot Spring, Iceland. *Clays Clay Miner.* **2002**, *50*, 174–185. [CrossRef]

54. Bristow, T.F.; Milliken, R.E. Terrestrial perspective on authigenic clay mineral production in ancient Martian lakes. *Clays Clay Miner.* **2011**, *59*, 339–358. [CrossRef]

55. Stumm, W.; Morgan, J.J. *Aquatic Chemistry An Introduction Emphasizing Chemical Equilibria in Natural Waters*; John Wiley & Sons: New York, NY, USA, 1981.

56. Stucki, J.W. Structural iron in smectites. In *Iron in Soils and Clay Minerals*; Springer: Dordrecht, The Netherlands, 1988; pp. 625–675.

57. Tuck, V.A.; Edyvean, R.G.J.; West, J.M.; Bateman, K.; Coombs, P.; Milodowski, A.E.; McKervey, J.A. Biologically induced clay formation in subsurface granitic environments. *J. Geochem. Explor.* **2006**, *90*, 123–133. [CrossRef]

58. Castanier, S.; Le Metayer-Levrel, G.; Perthuisot, J.P. Bacterial roles in the precipitation of carbonate minerals. In *Microbial Sediments*; Springer: Berlin/Heidelberg, Germany, 2000; pp. 32–39.

59. Braissant, O.; Decho, A.W.; Przekop, K.M.; Gallagher, K.L.; Glunk, C.; Dupraz, C.; Visscher, P.T. Characteristics and turnover of exopolymeric substances in a hypersaline microbial mat. *FEMS Microbiol. Ecol.* **2009**, *67*, 293–307. [CrossRef]

60. Dupraz, C.; Fowler, A.; Tobias, C.; Visscher, P.T. Stromatolitic knobs in Storr's Lake (San Salvador, Bahamas): A model system for formation and alteration of laminae. *Geobiology* **2013**, *11*, 527–548. [CrossRef]

61. Arp, G.; Thiel, V.; Reimer, A.; Michaelis, W.; Reitner, J. Biofilm exopolymers control microbialite formation at thermal springs discharging into the alkaline Pyramid Lake, Nevada, USA. *Sediment. Geol.* **1999**, *126*, 159–176. [CrossRef]

62. Sandford, P.A. Exocellular, microbial polysaccharides. In *Advances in Carbohydrate Chemistry and Biochemistry*; Academic Press: Cambridge, MA, USA, 1979; Volume 36, pp. 265–313.

63. Ueshima, M.; Tazaki, K. Possible role of microbial polysaccharides in nontronite formation. *Clays Clay Miner.* **2001**, *49*, 292–299. [CrossRef]

64. Abdel-Aziz, M.S.; Hamed, H.A.; Mouafi, F.E.; Gad, A.S. Acidic pH-shock induces the production of an exopolysaccharide by the fungus Mucor rouxii: Utilization of beet-molasses. *NY Sci. J.* **2012**, *5*, 52–61.

65. Mahapatra, S.; Banerjee, D. Fungal Exopolysaccharide: Production, Composition and Applications. *Microbiol. Insights* **2013**, *6*, 1–16. [CrossRef]

66. Wang, Y.; McNeil, B. Production of the fungal exopolysaccharide scleroglucan by cultivation of Sclerotium glucanicum in an airlift reactor with an external loop. *J. Chem. Technol. Biotechnol.* **1995**, *63*, 215–222. [CrossRef]

67. England, L.S.; Lee, H.; Trevors, J.T. Bacterial survival in soil: Effect of clays and protozoa. *Soil Boil. Biochem.* **1993**, *25*, 525–531. [CrossRef]

68. Stotzky, G.; Rem, L.T. Influence of clay minerals on microorganisms: I. Montmorillonite and kaolinite on bacteria. *Can. J. Microbiol.* **1966**, *12*, 547–563. [CrossRef]

69. Morley, G.F.; Gadd, G.M. Sorption of toxic metals by fungi and clay minerals. *Mycol. Res.* **1995**, *99*, 1429–1438. [CrossRef]

70. Gadd, G.M. Interactions of fungi with toxic metals. In *The Genus Aspergillus*; Springer: Boston, MA, USA, 1994; pp. 361–374.

71. Blaudez, D.; Jacob, C.; Turnau, K.; Colpaert, J.V.; Ahonen-Jonnarth, U.; Finlay, R.; Botton, B.; Chalot, M. Differential responses of ectomycorrhizal fungi to heavy metals in vitro. *Mycol. Res.* **2000**, *104*, 1366–1371. [CrossRef]

72. Perotto, S.; Martino, E. Molecular and cellular mechanisms of heavy metal tolerance in mycorrhizal fungi: What perspectives for bioremediation? *Min. Biotecnol.* **2001**, *13*, 55–63.

73. Baldrián, P. Interactions of heavy metals with white-rot fungi. *Enzym. Microb. Technol.* **2003**, *32*, 78–91. [CrossRef]

74. Gadd, G.M. Geomycology: biogeochemical transformations of rocks, minerals, metals and radionuclides by fungi, bioweathering and bioremediation. *Mycol. Res.* **2007**, *111*, 3–49. [CrossRef]

75. Stucki, J.W. Properties and behaviour of iron in clay minerals. *Dev. Clay Sci.* **2006**, *1*, 423–475.

76. Zhu, R.; Chen, Q.; Zhou, Q.; Xi, Y.; Zhu, J.; He, H. Adsorbents based on montmorillonite for contaminant removal from water: A review. *Appl. Clay Sci.* **2016**, *123*, 239–258. [CrossRef]

77. Yu, G.H.; He, P.J.; Shao, L.M. Characteristics of extracellular polymeric substances (EPS) fractions from excess sludges and their effects on bioflocculability. *Bioresour. Technol.* **2009**, *100*, 3193–3198. [CrossRef]

78. Kennedy, M.J.; Löhr, S.C.; Fraser, S.A.; Baruch, E.T. Direct evidence for organic carbon preservation as clay-organic nanocomposites in a Devonian black shale; from deposition to diagenesis. *Earth Planet. Sci. Lett.* **2014**, *388*, 59–70. [CrossRef]

79. Playter, T.; Konhauser, K.; Owttrim, G.; Hodgson, C.; Warchola, T.; Mloszewska, A.M.; Sutherland, B.; Bekker, A.; Zonneveld, J.-P.; Pemberton, S.G.; et al. Microbe-clay interactions as a mechanism for the preservation of organic matter and trace metal biosignatures in black shales. *Chem. Geol.* **2017**, *459*, 75–90. [CrossRef]

80. LaLonde, K.; Mucci, A.; Ouellet, A.; Gelinas, Y. Preservation of organic matter in sediments promoted by iron. *Nature* **2012**, *483*, 198–200. [CrossRef]

81. Parenteau, M.N.; Jahnke, L.L.; Farmer, J.D.; Cady, S.L. Production and Early Preservation of Lipid Biomarkers in Iron Hot Springs. *Astrobiology* **2014**, *14*, 502–521. [CrossRef]

82. Summons, R.E.; Amend, J.P.; Bish, D.; Buick, R.; Cody, G.D.; Marais, D.J.D.; Dromart, G.; Eigenbrode, J.L.; Knoll, A.H.; Sumner, D.Y. Preservation of Martian Organic and Environmental Records: Final Report of the Mars Biosignature Working Group. *Astrobiology* **2011**, *11*, 157–181. [CrossRef]

83. Magnabosco, C.; Lin, L.-H.; Dong, H.; Bomberg, M.; Ghiorse, W.; Stan-Lotter, H.; Pedersen, K.; Kieft, T.L.; Van Heerden, E.; Onstott, T.C. The biomass and biodiversity of the continental subsurface. *Nat. Geosci.* **2018**, *11*, 707. [CrossRef]

84. Beaufort, D.; Rigault, C.; Billon, S.; Billault, V.; Inoue, A.; Inoue, S.; Patrier, P. Chlorite and chloritization processes through mixed-layer mineral series in low-temperature geological systems—a review. *Clay Miner.* **2015**, *50*, 497–523. [CrossRef]

85. Canfield, D.E. A new model for Proterozoic ocean chemistry. *Nature* **1998**, *396*, 450–453. [CrossRef]

86. Tarduno, J.A.; Duncan, R.A.; Scholl, D.W. *Leg 197 Summary: Proceedings of the Ocean Drilling Program, Initial Reports*; Ocean Drilling Program Texas A&M University: College Station TX, USA, 2002; Volume 197, pp. 1–92.

87. Orcutt, B.N.; Wheat, C.G.; Rouxel, O.; Hulme, S.; Edwards, K.J.; Bach, W. Oxygen consumption rates in subseafloor basaltic crust derived from a reaction transport model. *Nat. Commun.* **2013**, *4*, 2539. [CrossRef]

88. Bach, W.; Edwards, K.J. Iron and sulfide oxidation within the basaltic ocean crust: implications for chemolithoautotrophic microbial biomass production. *Geochim. Cosmochim. Acta* **2003**, *67*, 3871–3887. [CrossRef]

89. Tosca, N.J.; Knoll, A.H. Juvenile chemical sediments and the long term persistence of water at the surface of Mars. *Earth Planet. Sci. Lett.* **2009**, *286*, 379–386. [CrossRef]

90. Dong, H.; Jaisi, D.P.; Kim, J.; Zhang, G. Microbe-clay mineral interactions. *Am. Mineral.* **2009**, *94*, 1505–1519. [CrossRef]

91. Sakakibara, M.; Sugawara, H.; Tsuji, T.; Ikehara, M. Filamentous microbial fossil from low-grade metamorphosed basalt in northern Chichibu belt, central Shikoku, Japan. *Planet. Space Sci.* **2014**, *95*, 84–93. [CrossRef]

MDPI

St. Alban-Anlage 66

4052 Basel

Switzerland

Tel. +41 61 683 77 34

Fax +41 61 302 89 18

www.mdpi.com

Geosciences Editorial Office

E-mail: geosciences@mdpi.com

www.mdpi.com/journal/geosciences